2-17

Lecture Notes in Statistics

Edited by J. Berger, S. Fienberg, J. Gani, K. Krickeberg, and B. Singer

54

Kirti R. Shah
Bikas K. Sinha

D1606717

Theory of Optimal Designs

Springer-Verlag
New York Berlin Heidelberg London Paris Tokyo Hong Kong

Authors

Kirti R. Shah
Department of Statistics and Actuarial Science
University of Waterloo
Waterloo, Ontario, Canada N2L 3G1

Bikas K. Sinha
Stat-Math Division, Indian Statistical Institute
203 B.T. Road, Calcutta - 700035, India

Mathematical Subject Classification Code: 62K05

ISBN 0-387-96991-8 Springer-Verlag New York Berlin Heidelberg
ISBN 3-540-96991-8 Springer-Verlag Berlin Heidelberg New York

Printed in Germany

Printing and binding: Druckhaus Beltz, Hemsbach/Bergstr.
2847/3140-543210 – Printed on acid-free paper

DEDICATED

TO THE MEMORY

OF

LATE PROFESSOR JACK KIEFER

WHOSE FUNDAMENTAL CONTRIBUTIONS

HAVE BEEN A SOURCE

OF

ILLUMINATION AND INSPIRATION

TO OUR THOUGHTS ON THIS TOPIC

PREFACE

There has been an enormous growth in recent years in the literature on discrete optimal designs. The optimality problems have been formulated in various models arising in the experimental designs and substantial progress has been made towards solving some of these. The subject has now reached a stage of completeness which calls for a self-contained monograph on this topic. The aim of this monograph is to present the state of the art and to focus on more recent advances in this rapidly developing area.

We start with a discussion of statistical optimality criteria in Chapter One. Chapters Two and Three deal with optimal block designs. Row-column designs are dealt with in Chapter Four. In Chapter Five we deal with optimal designs with mixed effects models. Repeated measurement designs are considered in Chapter Six. Chapter Seven deals with some special situations and Weighing designs are discussed in Chapter Eight.

We have endeavoured to include all the major developments that have taken place in the last three decades. The book should be of use to research workers in several areas including combinatorics as well as to the experimenters in diverse fields of applications. Since the details of the construction of the designs are available in excellent books, we have only pointed out the designs which have optimality properties. We believe, this will be adequate for the experimenters.

The variety of research areas involving optimality considerations is reflected in the wide spectrum of the contributions made by researchers all over the world. Through the medium of this monograph, we express our appreciation of these contributions towards the furtherance of this fascinating subject. We hope that this monograph will stimulate further research in this area. Finally, we would like to apologize to various authors whose work has been missed partly by oversight and partly due to constraints of time and space.

ACKNOWLEDGEMENTS

We would like to thank Professor A.C. Mukhopadhyaya and Dr. (Mrs.) S. Bagchi (formerly Mukhopadhyaya) for their constructive comments and suggestions on an earlier draft of this monograph.

We are especially grateful to Mrs. Lynda Clarke at the University of Waterloo for her meticulous typing of this monograph. We would also like to thank Mrs. Janice Gaddy of North Carolina State University for typing of an earlier draft and Ms. Lucy Roefs for typing of some parts of this monograph.

We are also indebted to the Department of Statistics and Actuarial Science at the University of Waterloo for providing all the needed facilities and to the National Sciences and Engineering Research Council of Canada for providing support for the work on this monograph.

Kirti R. Shah Bikas K. Sinha

To Professors M.C. Chakrabarti, C.R. Rao and J. Roy for the encouragement and inspiration they have given me. They also guided me during my early excursions into this newly emerging area.

To my wife Daksha for her patience and support during the long hours devoted to the preparation of this manuscript.

Kirti R. Shah

To my revered teacher the late Professor H.K. Nandi under whose inspiration and guidance I began work on optimality nearly twenty years back.

To Professor J.K. Ghosh for his encouragement and sincere interest in this monograph.

To my wife Pritha who has relieved me of much of the tedious but necessary trivia that encumber the lives of most family men, and ensured that I could devote myself whole-heartedly to my work throughout my professional career.

Bikas K. Sinha

TABLE OF CONTENTS

CHAPTER ONE

OPTIMALITY CRITERIA IN DESIGN OF EXPERIMENTS

1. General Objectives

Experimentation is an essential part of any problem of decision-making. Whenever one is faced with the necessity of accepting one out of a set of alternative decisions, one has to undertake some experiments to collect observations on which the decision has to be based. In order that it may be possible to select an *optimum* decision procedure, the choice of the experiment must also be *optimum* in some sense. This is how the problem of *optimal designing of experiments* arises.

The concept of *sufficiency* in comparing statistical experiments is well-known. (Blackwell (1951, 1953), Blackwell and Girshick (1954)). Roughly speaking, an experiment $\mathcal{E}^*$ resulting in a random variable U having law of distribution $\{F_\theta(.), \theta \epsilon \Omega\}$ is said to be sufficient for another experiment $\mathcal{E}$ resulting in a random variable V having law of distribution $\{G_\theta(.), \theta \epsilon \Omega\}$ if given an observation u on U, it is possible to generate an observation v on V using a *known* randomization procedure i.e., using an observation w of another random variable W having a law of distribution $Z(.|u)$ which is completely *known* (and, certainly, independent of θ). If the above holds, we say that U is sufficient for V and write $U \gg V$. Clearly, when $U \gg V$, it is enough to observe U to generate V, if needed. Moreover, it is known (vide Blackwell and Girshick (1954)) that when $U \gg V$, for any decision rule $\delta(V)$ for θ based on V, there exists a corresponding decision rule $\delta^*(U)$ for θ based on U which is at least as good (in the sense of having equal or smaller *risk*).

In case of experimental designs fitting into the usual ANOVA (Analysis of Variance) model, the existence of a sufficient experimental design in the above sense would certainly settle the problem of choice of the best experimental design. However, it is easy to verify that such sufficient designs do *not* exist even in the simplest set-ups. In such design set-ups, therefore, the choice of an optimum experimental design is certainly governed by what the term *optimum* connotes. The theory of optimal experimental designs centers around the problems of *characterization* and *construction* of designs which are *optimum* in some sense from among a number of alternative experiments in a given design set-up.

2. The Linear Model Set-up

For the most part, we will assume that the observation-vector $\mathbf{Y}(n\times 1)$ follows a standard linear model. Specifically, we may write

$$E(\mathbf{Y}) = \mathbf{X}\boldsymbol{\theta}\,,\ Cov(\mathbf{Y}) = \sigma^2\mathbf{I}_n \qquad (1.2.1)$$

where $\mathbf{X}$ is an $n\times t$ matrix of known coefficients (usually known as the *design matrix*) and $\boldsymbol{\theta}$ is a $t\times 1$ vector of unknown parameters. Here σ^2 is the constant error variance (usually unknown) and $\mathbf{I}_n$ is the identity matrix of order n. We will refer to (1.2.1) as a *fixed-effects* linear model.

Suppose we are interested in a component $\boldsymbol{\theta}_1$ of $\boldsymbol{\theta}$. We write $\boldsymbol{\theta}' = (\boldsymbol{\theta}_1':\boldsymbol{\theta}_2')$ and accordingly partition $\mathbf{X}$ as $\mathbf{X} = (\mathbf{X}_1:\mathbf{X}_2)$ so that the model becomes

$$E(\mathbf{Y}) = \mathbf{X}_1\boldsymbol{\theta}_1 + \mathbf{X}_2\boldsymbol{\theta}_2, \; Cov(\mathbf{Y}) = \sigma^2\mathbf{I}_n. \tag{1.2.2}$$

Under the assumption of normality of the errors, the (Fisher) information matrix of $\boldsymbol{\theta}_1$ under the above model is given by (apart from the multiplier of σ^{-2})

$$\mathbf{I}(\boldsymbol{\theta}_1) = \mathbf{X}_1'\mathbf{X}_1 - \mathbf{X}_1'\mathbf{X}_2(\mathbf{X}_2'\mathbf{X}_2)^-\mathbf{X}_2'\mathbf{X}_1 = \mathbf{X}_1'(\mathbf{I}-\mathbf{P}_{\mathbf{X}_2})\mathbf{X}_1 \tag{1.2.3}$$

where $\mathbf{A}^-$ denotes a generalized inverse (g-inverse) of $\mathbf{A}$ and $\mathbf{P}_\mathbf{A}$ denotes the orthogonal projection matrix on the column space of the matrix $\mathbf{A}$.

Clearly, $\mathbf{I}(\boldsymbol{\theta}) = \mathbf{X}'\mathbf{X}$. In general, $\mathbf{I}(\boldsymbol{\theta}_1)$ is non-negative definite (nnd) and it is positive definite if and only if (iff) *all* components of $\boldsymbol{\theta}_1$ are *estimable*.

Following Kiefer (1958), an information matrix will be said to be *completely symmetric* (c.s.) iff all its diagonal elements are equal to one another *and* so also are its off-diagonal elements.

In a *block design* set-up, we are primarily interested in drawing inferences about the treatment effects which may be identified with the components of $\boldsymbol{\theta}_1$ in the above model.

If $\mathbf{N}$ denotes the *incidence* matrix of order $v \times b$ of a block design involving v treatments applied in b blocks each of size k, then the information matrix for the treatment effects, usually denoted by $\mathbf{C}$, has the representation

$$\mathbf{C} = \mathbf{r}^\delta - k^{-1}\mathbf{N}\mathbf{N}' \tag{1.2.4}$$

where $\mathbf{r} = \mathbf{N}\mathbf{1}$, $\mathbf{r}^\delta = Diag(r_1, \ldots, r_v)$ and $\mathbf{1} = (11 \cdots 1)'$ of appropriate order. In such a set-up, a linear combination of treatment effects with coefficient vector $\mathbf{p}$ is *estimable* iff $\mathbf{p}$ belongs to the column-space of $\mathbf{C}$. Since $\mathbf{C}\mathbf{1} = \mathbf{0}$, only the linear functions $\mathbf{p}'\boldsymbol{\theta}_1$ with coefficient vector $\mathbf{p}$ satisfying $\mathbf{p}'\mathbf{1} = 0$ can be estimated. Such linear functions are referred to as *treatment contrasts*. It is known that rank($\mathbf{C}$) = $v-1$ iff *all* treatment contrasts are estimable and in that case, the underlying design is said to be *connected*. There is an equivalent notion of connectedness referring to what (the late Professor Raj Chandra) Bose called the "chain definition" of connectedness. We refer to Chakrabarti (1962) for further details.

In the sequel, we will be *exclusively* dealing with connected designs.

In an experimental set-up comprising of bk experimental units divided into b groups each of size k, a block design seeks to eliminate suspected heterogeneity among the groups. On the other hand, if one suspects two sources of heterogeneity, one may choose bk experimental units which can be arranged in a $k \times b$ array in such a way that the heterogeneity occurs across the rows and across the columns. In this case, a design involving v treatments will consist of an arrangement of treatments to the bk units of the $k \times b$ array. Such designs are termed *row-column* designs. Here we come across two types of incidence matrices: treatments *vs* rows and treatments *vs* columns, to be denoted by $\mathbf{M}(v\times k)$ and $\mathbf{N}(v\times b)$ respectively. Here the information matrix for treatment effects turns out to be

$$\mathbf{C} = \mathbf{r}^{\delta}-k^{-1}\mathbf{NN}'-b^{-1}\mathbf{MM}'+b^{-1}k^{-1}\mathbf{rr}' \tag{1.2.5}$$

where $\mathbf{r} = \mathbf{N1} = \mathbf{M1}$, $\mathbf{r}^{\delta} = Diag(r_1, \ldots, r_v)$. We refer to Shrikhande (1951) for the derivation of the above expression. In a row-column design as well, *all* treatment contrasts are *estimable* iff the resulting $\mathbf{C}$-matrix is of rank $(v-1)$ in which case the design is said to be connected. This is because the $\mathbf{C}$-matrix continues to have the property $\mathbf{C1} = \mathbf{0}$.

It should be remarked that a disconnected block design has a simple structure which is easy to recognize by utilizing the chain definition of connectedness. However, this is not true for a row-column design because a row-column design can fail to be connected even when each of the two component block designs is connected (Shah and Khatri (1973)). There does not appear to be a simple way of checking for connectedness of a row-column design.

It is quite possible that there is *no* heterogeneity in the n experimental units which are being utilized for the purpose of treatment comparisons. This gives rise to what is popularly known as a *Completely Randomized Design* (**CRD**). Clearly, in a **CRD**, the designing problem corresponds to the problem of allocation of the experimental units among various treatments. In this case, with n experimental units and v treatments having replication numbers $n_1, \ldots, n_v$ such that $n_i \geq 1$, $1 \leq i \leq v$, $\sum_1^v n_i = n$, the $\mathbf{C}$-matrix assumes the form

$$\mathbf{C} = \mathbf{n}^{\delta}-\frac{\mathbf{nn}'}{n}$$

where $n = \sum_1^v n_i$, $\mathbf{n} = (n_1, \ldots, n_v)'$ and $\mathbf{n}^{\delta} = Diag(n_1, n_2, \ldots, n_v)$. We will *not* study the optimality aspects of **CRD**s in this monograph. This will be left as an exercise. Instead, we will have occasion to deal with other design set-ups in later Chapters.

3. Choice of Optimality Criteria

3.1 Basic Considerations

We address the problem of suggesting a suitable class of optimality criteria for comparison of experiments involving linear models. Specifically, denote by $\mathcal{C}$ the class of available $\mathbf{C}$-matrices in a given experimental set-up say, for example, the traditional block design set-up or a row-column design set-up where the interest lies primarily in the so-called fixed treatment effects. Denote by Φ the class of optimality functionals Φ defined on the members of $\mathcal{C}$.

The optimality functionals may be aimed at providing symmetric measures of the *lack of information* contained in the $\mathbf{C}$-matrices (which essentially represent information matrices for the treatment effects) and, hence, it would be reasonable to impose invariance w.r.t. the symmetric group of permutations on the treatment symbols. Let $g \epsilon S_v$, the symmetric group of permutations of order v, and let $\mathbf{G}_g$ denote the corresponding permutation matrix i.e., the matrix obtained by applying g to the columns of the identity matrix. For a given g, let $\mathbf{C}_g = \mathbf{G}_g'\mathbf{C}\mathbf{G}_g$. It is easy to see that $\mathbf{C}_g$ corresponds to a variant of the design based on the permutation g applied to the set of treatments.

Then the first reasonable requirement to be satisfied by the optimality functionals Φ may be laid down as

(i) $\Phi(\mathbf{C}) = \Phi(\mathbf{C}_g)$ for every member g of the symmetric group of permutations.

The next requirement is in terms of comparing the information matrices $\mathbf{C}_1$ and $\mathbf{C}_2$ arising out of two different (non-isomorphic) designs. If $\mathbf{C}_1 \geq \mathbf{C}_2$ (in the sense of $(\mathbf{C}_1 - \mathbf{C}_2)$ being nnd), then it would be reasonable to require that the Φ-measure of the lack of information in $\mathbf{C}_1$ is less than that in $\mathbf{C}_2$ i.e.,

(ii) $\mathbf{C}_1 \geq \mathbf{C}_2 => \Phi(\mathbf{C}_1) \leq \Phi(\mathbf{C}_2)$.

Another requirement may be based on our understanding of repeated use of better designs. Thus, specifically, we require that $\mathbf{C}_2$ is *Φ-better* than $\mathbf{C}_1$ iff so are t copies of the former design compared to t copies of the latter. Mathematically, we stipulate the following:

(iii) $\Phi(\mathbf{C}_1) \geq \Phi(\mathbf{C}_2) \Leftrightarrow \Phi(t\mathbf{C}_1) \geq \Phi(t\mathbf{C}_2)$ for all positive integers $t \geq 1$.

Finally, we may wish to compare two *derived* designs. Given any design d to start with, we construct d_g as a variation of d by permuting the treatments according to the permutation g. Clearly, $\mathbf{G}_g'\mathbf{C}\mathbf{G}_g = \mathbf{C}_g$ is the resulting $\mathbf{C}$-matrix of d_g. Then $\Sigma t_g \mathbf{G}_g'\mathbf{C}\mathbf{G}_g$ represents the $\mathbf{C}$-matrix underlying the design formed by considering t_g copies of d_g, simultaneously for every $t_g > 0$. On the other hand, $(\Sigma t_g)\mathbf{C}$ corresponds to the $\mathbf{C}$-matrix of another *derived* design which is simply formed of (Σt_g) copies of the original design d. Since the optimality functionals are permutation-invariant (the requirement (i)), one would expect that a combination of various forms of d would do rather better than an exclusive use of d itself. This leads us to the requirement

(iv) $\Phi(\Sigma t_g \mathbf{G}_g'\mathbf{C}\mathbf{G}_g) \leq \Phi((\Sigma t_g)\mathbf{C})$ for all non-negative integers t_g's, at least one t_g being positive.

In particular, (iv) implies

(iv)′ $\Phi(\Sigma(\mathbf{G}_g'\mathbf{C}\mathbf{G}_g)) \leq \Phi((v!)\mathbf{C})$.

We may say that a design is *optimal* in a very general sense if the underlying $\mathbf{C}$-matrix minimizes each optimality functional Φ (in the class $\boldsymbol{\Phi}$) satisfying the requirements (i) - (iv).

We may refer to (iv)′ as *symmetry* requirement on the optimality functionals Φ. The requirement (iv) may be referred to as the property of *weak convexity*. This is motivated by the fact that (i) and (iii) together with the usual notion of convexity (described below in (iv)″ of subsection 3.2) implies (iv). In fact, (iv) is satisfied by all functionals of the form $\Phi(\mathbf{C}) = g(f(\mathbf{C}))$ where g is monotone increasing, and f is convex in the *usual* sense satisfying (i) and (iii). Certainly plenty of examples of such forms of Φ can be constructed where Φ itself is *not* convex. The following example, communicated by (the late Professor) Kiefer (in a private communication), is one of them. Writing tr for the trace of a matrix, let

$$\Phi(\mathbf{C}) = \{tr(\mathbf{C}^2)\}^{1/4}$$

Taking $\mathbf{C}_1 = 9(\mathbf{I}_v - \mathbf{1}\mathbf{1}'/v)$ and $\mathbf{C}_2 = 16(\mathbf{I}_v - \mathbf{1}\mathbf{1}'/v)$, it is seen that

$$\Phi(\frac{1}{2}(\mathbf{C}_1+\mathbf{C}_2)) > \frac{1}{2}[\Phi(\mathbf{C}_1) + \Phi(\mathbf{C}_2)]$$

thereby violating convexity in respect of Φ. However, since Φ is a monotone increasing function of $tr(\mathbf{C}^2)$ which is a convex function of $\mathbf{C}$, weak convexity holds. The fact that (iv)′ is weaker than (iv) can be seen through the following example: Consider two block designs each with $b = k = 2$, $v = 3$. The rows are blocks.

d_1

1	2
1	3

d_2

2	1
2	3

Consider the usual fixed-effects additive model without interaction. Then

$$2\mathbf{C}_1 = \begin{pmatrix} 2 & -1 & -1 \\ -1 & 1 & 0 \\ -1 & 0 & 1 \end{pmatrix}, \quad 2\mathbf{C}_2 = \begin{pmatrix} 1 & -1 & 0 \\ -1 & 2 & -1 \\ 0 & -1 & 1 \end{pmatrix}, \quad \mathbf{C}_2 = \mathbf{G}_g'\mathbf{C}_1\mathbf{G}_g \text{ with } g = (1 \Leftrightarrow 2).$$

Define $\Phi(\mathbf{C}) = tr(\mathbf{C}) + \{x_{(v-1)} - x_{(1)}\}^{\delta}\{\frac{A}{x_{(v-1)}} + \frac{1}{x_{(1)}}\}$ for some $\delta > 0$, $A > 0$ where $x_{(1)}$ and $x_{(v-1)}$ denote the smallest and the largest non-zero eigenvalues of $\mathbf{C}$. It now turns out that for $\delta = 0.1$ and $A = 10$, $\Phi(\mathbf{C}_1+\mathbf{C}_2) > \Phi(2\mathbf{C}_1)$ while it is easy to verify that $\Phi(\sum_g \mathbf{G}_g'\mathbf{C}_1\mathbf{G}_g) < \Phi(6\mathbf{C}_1)$ for *any* choice of $\delta > 0$, $A > 0$.

When we assume (i), (ii) and (iii), the following are seen to be progressively weaker:

(1) Φ is convex in the usual sense

(2) Φ is monotone increasing function of a convex function

(3) Φ is weakly convex in the sense of (iv)

(4) Φ satisfies the symmetry requirement (iv)′.

We may summarize the above observations by stating that

$$\Phi_C \subseteq \Phi_{IC} \subseteq \Phi_{WC} \subseteq \Phi_S$$

where

Φ_C = class of convex functionals

= {Φ: Φ satisfies (*i*), (*ii*), (*iii*), (*iv*)''};

Φ_{IC} = class of increasing functions of convex functionals

= {Φ: $\Phi(\mathbf{C}) = g(f(\mathbf{C}))$, $g \uparrow$ and f satisfies (*i*), (*ii*), (*iii*), (*iv*)''};

Φ_{WC} = class of weakly convex functionals

= {Φ: satisfies (*i*), (*ii*), (*iii*),(*iv*)};

Φ_S = class of symmetric functionals

= {Φ: Φ satisfies (*i*), (*ii*), (*iii*), (*iv*)'}.

In the above, (iv)'' corresponds to the usual definition of convexity stated below.

3.2 Notion of Universal Optimality

Kiefer (1975) introduced the notion of *Universal Optimality* in the following manner. Consider optimality functionals Φ defined on the set of *all* C-matrices which satisfy (i), a version of (ii) namely,

(ii)' $\Phi(t\mathbf{C})$ is non-increasing in t, $t \geq 0$

and the condition of convexity in the usual sense i.e.

(iv)'' $\Phi(\alpha\mathbf{C}_1 + (1-\alpha)\mathbf{C}_2) \leq \alpha\Phi(\mathbf{C}_1) + (1-\alpha)\Phi(\mathbf{C}_2)$,

for $0<\alpha<1$ and for *any* pair of C-matrices $\mathbf{C}_1$ and $\mathbf{C}_2$. If a design is optimal w.r.t. all such optimality functionals Φ, it is said to be universally optimal.

We shall try to examine this notion of universal optimality in the light of the basic requirements set forth in subsection 3.1. The *essential* difference is in the condition of convexity (iv)'' imposed on the optimality functionals which certainly implies weak convexity (iv) and hence the symmetry condition (iv)'. However, usual convexity does *not* appear to have any statistical interpretation in this context. On the other hand, condition (iv)' (or even condition (iv) of which (iv)' is a special case) is appealing in a reasonable statistical sense. Further, (iv) or (iv)' relates only to a set of feasible C-matrices which correspond to actual designs whereas the convexity condition in the usual sense will generally relate to non-feasible C-matrices, when considering arbitrary convex combinations $\alpha\mathbf{C}_1+(1-\alpha)\mathbf{C}_2$.

In view of the above, we may adopt the following definition of extended universal optimality.

Extended Universal Optimality. A design will be said to be universally optimal in an *extended* sense in a given design class if the underlying C-matrix minimizes every optimality functional Φ satisfying (i), (ii), (iii) and (iv)′ among all C-matrices for designs in that class.

Kiefer's original proof of universal optimality of *balanced* designs requires the convexity condition in the usual sense. However, analogous arguments enable us to provide a neat proof of the following modified version of Proposition 1 of Kiefer (1975).

Proposition 1 (Restated). If there is a feasible C-matrix which is completely symmetric and has maximum trace, then the underlying design is universally optimal in the extended sense of minimizing Φ *simultaneously* for all functionals Φ satisfying (i), (ii), (iii) and (iv)′.

Proof. Suppose a feasible C-matrix $\mathbf{C}_0$ satisfies the conditions laid down above. Then $\mathbf{C}_0 = a_0(\mathbf{I}_v - v^{-1}\mathbf{1}\mathbf{1}')$ where $a_0 = tr\,(\mathbf{C}_0)/(v-1)$. Let C be the C-matrix for any competing design in the same class. It is easy to verify that $\sum_g \mathbf{C}_g$ has the representation $\sum_g \mathbf{C}_g = a(\mathbf{I}_v - v^{-1}\mathbf{1}\mathbf{1}')$ where $a = v!tr\,(\mathbf{C})/(v-1)$. Since $\mathbf{C}_0$ possesses maximum trace, $v!a_0 > a$. Further, as $\sum_g \mathbf{C}_g$ and $\mathbf{C}_0$ are both multiples of $(\mathbf{I}_v - v^{-1}\mathbf{1}\mathbf{1}')$, we get

$$\sum \mathbf{C}_g = a(\mathbf{I}_v - v^{-1}\mathbf{1}\mathbf{1}') < v!a_0(I_v - v^{-1}\mathbf{1}\mathbf{1}') = v!\mathbf{C}_0$$

Hence, by (ii),

$$\Phi(\sum_g \mathbf{C}_g) \geq \Phi(v!\mathbf{C}_0)$$

Now suppose $\Phi(\mathbf{C}) < \Phi(\mathbf{C}_0)$. Then using (iv)′ and (iii), we get

$$\Phi(\sum_g \mathbf{C}_g) \leq \Phi(v!\mathbf{C}) < \Phi(v!\mathbf{C}_0)$$

which is a contradiction. This completes the proof.*

Recently, Yeh (1986) suggested another set of sufficient conditions for universal optimality in Kiefer's sense. In the spirit of the notion of extended universal optimality discussed above, we may state Yeh's result in the following form.

Let $\mathbf{C}^*$ be a feasible C-matrix such that for every feasible competitor C, there are non-negative integers a_g, $g \in \mathbf{S}_v$, $1 \leq \sum a_g < \infty$ for which $(\sum a_g)\mathbf{C}^* = \sum_g a_g \mathbf{C}_g$. Then C^* is optimal w.r.t. every functional Φ which satisfies (i), (ii), (iii) and (iv).

*As Kiefer (1975) has observed, $-tr(\mathbf{C})$ is a functional satisfying (i), (ii), (iii) and (iv)′. Consequently, a **necessary** condition for a design to be universally optimal is that the corresponding C-matrix possesses maximum trace among all competing C-matrices.

Yeh (1986, 1988) discussed interesting (but limited) applications of this result in the characterization and construction of such optimal designs in a suitable subclass of designs when $k = \pm 1 (mod\ v)$.

In the next two subsections we shall present various widely used optimality criteria discussed in the literature. In the final subsection, we shall discuss the relationships among them.

3.3 Specific Optimality Criteria

Historically speaking, Smith (1918) appears to be the first to formally introduce a specific *optimality criterion* in comparing designs in a given experimental set-up. Smith (1918) dealt with *regression designs* and suggested a response function criterion.

Much later, Wald (1943) introduced two important optimality criteria in the context of designs eliminating heterogeneity in two directions. On the other hand, the notion of *Efficiency Factor* seems to have been in use by the experimenters soon after incomplete block designs were introduced by Fisher and Yates. (See Kempthorne (1956), for example). However, it was Kiefer (1958) who, for the first time, formalized the above optimality criteria in a general set-up and in a unified way. (To differentiate the various criteria discussed by him, he attributed meaningful *names* to these criteria. These have been universally accepted.) Since that time some further generalizations of the above criteria as also a few others have been proposed. We will first present the *Generalized Criteria* which cover most of the criteria currently in use. These were introduced by Kiefer (1975) and studied in detail by Cheng (1978). Then we will specialize to the specific ones and discuss their statistical interpretations as far as applicable. In subsection 3.4, we mention a few other criteria which are indeed statistically meaningful but are *not* included in the above class. Following Cheng (1978), let us consider a class of optimality functionals (to be minimized) which are of the form

$$\psi_f(\mathbf{C}) = \sum_{1}^{v-1} f(x_i) \tag{1.3.1}$$

where f is defined over $(0,M_0)$ and x_i's are positive eigenvalues of $\mathbf{C}$. Here, $M_0 = \max tr(\mathbf{C})$ where the maximum is taken over the class of relevant designs under consideration and $tr(\mathbf{C})$ denotes the trace of the $\mathbf{C}$-matrix underlying the design. The function f in (1.3.1) is assumed to satisfy the following conditions:

$$\left.\begin{array}{l}\text{(i) } f \text{ is continuously differentiable on } (0,M_0)\\ \text{(ii) } f'<0, f''>0, f'''<0 \text{ on } (0,M_0)\\ \text{(iii) } f \text{ is continuous at } 0 \text{ and } f(0) = \lim\limits_{x \to 0+} f(x) = +\infty.\end{array}\right\} \tag{1.3.2}$$

(This last condition on f ensures that the designs having eigenvalue(s) near zero *cannot* be optimal).

This class may be *extended* by including *all* monotone functions of ψ. A further extension is also possible, by including point-wise limits of suitable sequences of optimality functionals so obtained.

The optimality functionals ψ in (1.3.1) together with their extensions described above will be termed *Generalized Optimality Criteria*.

Cheng (1978) also considered another class of optimality functionals where the condition (ii) on f in (1.3.2) is replaced by

$$\text{(ii)}' \quad f'<0,\ f''>0,\ f'''>0 \text{ on } (0, M_0).$$

This latter class does *not* yield any statistically meaningful criterion and, hence, will *not* be dealt with any further.

The most widely used optimality criteria, called **A**- and **D-optimality** criteria, are of the type (1.3.1). Specifically, these are defined as follows:

A-optimality Minimize $\sum 1/x_i$ over the entire class (1.3.3)

$(f(x) = \frac{1}{x})$

D-optimality Minimize $-\sum \log x_i$ over the entire class

$(f(x) = -\log x)$ or equivalently (1.3.4)

Minimize Πx_i^{-1} over the entire class

Two further criteria suggested in the literature (Eccleston and Hedayat (1974), Shah (1960)) are also of the above type and are given by

S-optimality Minimize Σx_i^2 over the entire class

(M-S)-optimality Minimize Σx_i^2 over the subclass $\mathcal{C}'$(of $\mathcal{C}$)

defined by $\mathcal{C}'$ = {**C**-matrices for designs which maximize Σx_i}

The following class of optimality criteria introduced by Kiefer (1975) and defined as

$$\{\Phi_p(\mathbf{C}) = (\Sigma x_i^{-p}/(v-1))^{1/p},\ 0<p<\infty\} \tag{1.3.5}$$

can be viewed as an extension of (1.3.1). The class (1.3.5) generates what are known as Φ_p-optimality criteria. It is clear that **A**-optimality criterion is a Φ_p-criterion with $p = 1$. It can be shown that **D**-optimality criterion is a point-wise limit of Φ_p-criteria as $p \to 0$. Another widely used criterion is obtained by taking a point-wise limit of Φ_p-criteria as $p \to \infty$ and this yields $\lim_{p\to\infty} \Phi_p(\mathbf{C}) = \max_i(\frac{1}{x_i})$. The resulting criterion is restated in the well-known form of

E-optimality Maximize the smallest positive eigenvalue of **C** (1.3.6)

It can be easily verified that if a design has a completely symmetric **C**-matrix and is Φ_p-optimal, then it is Φ_q-optimal for any $q>p$. Thus, in particular, **D**-optimality (**A**-optimality) implies **A**- and **E**-optimality (respectively, **E**-optimality) whenever such an optimal design possesses a c.s. **C**-matrix. We have presented the notions of **A**–, **D**– and **E**–optimality criteria in a unified way through the Φ_p-criterion. However, these have been widely in use well before such a formulation was given by Kiefer (1958, 1975) and Cheng (1978). This is because these arise quite naturally in various statistical multiparameter inference problems. To see this, let $\boldsymbol{\theta} = (\theta_1, \theta_2, \ldots, \theta_p)'$ be an unknown parameter vector involved in the distribution of a random vector **Y**. Suppose $\hat{\boldsymbol{\theta}}$ refers to an estimate of $\boldsymbol{\theta}$ having a positive definite dispersion matrix $\sigma^2\Sigma$ where $\sigma>0$ is unknown and Σ is determined through the design resulting in the random vector **Y**.

The idea of **E**-optimality arises in the following situations:

(i) Suppose we want to estimate *all normed* linear functions of the components of $\boldsymbol{\theta}$ i.e., $\mathbf{a}'\boldsymbol{\theta}$ subject to $\mathbf{a}'\mathbf{a} = 1$. Then the maximum variance of such a linear function is $\lambda_{\max}(\Sigma)$ i.e., the maximum eigenvalue of Σ and we might wish to minimize this by a proper choice of the design resulting in the random vector **Y**.

(ii) Suppose we want to test the hypothesis H_0: $\boldsymbol{\theta} = \mathbf{0}$ against $\boldsymbol{\theta} \neq \mathbf{0}$. Assuming normality of $\hat{\boldsymbol{\theta}}$, the power function of the F-test can be studied through the noncentrality parameter $\boldsymbol{\theta}'\Sigma^{-1}\boldsymbol{\theta}$ and we might wish to maximize the minimum power over the contours given by $\boldsymbol{\theta}'\boldsymbol{\theta} = c(>0)$. This again should be done by a proper choice of the design resulting in the random vector **Y**.

In both cases, the optimality criterion comes out as

Minimize the maximum eigenvalue of Σ (1.3.7)

Next we explain the notion of **D**-optimality. Suppose we are interested in the $100(1-\alpha)\%$ confidence ellipsoid for $\boldsymbol{\theta}$ assuming normality of $\hat{\boldsymbol{\theta}}$. Clearly, the objective would be to minimize the volume of the ellipsoid by a proper choice of the underlying design. It is known that the volume of this ellipsoid is proportional to the square root of the determinant of Σ. The determinant of Σ, also known as the generalized variance of $\hat{\boldsymbol{\theta}}$, is given by the product of the eigenvalues of Σ. This leads to the criterion

Minimize the product of the eigenvalues of Σ (1.3.8)

Finally, as to the notion of **A**-optimality, the idea is to minimize the average variance of the components of $\hat{\boldsymbol{\theta}}$. Stated equivalently, this criterion is

Minimize the sum of the eigenvalues of Σ (1.3.9)

To connect these general notions of **A**–, **D**– and **E**– optimality as embodied in (1.3.7) – (1.3.9) to those described earlier in the set-up of experimental designs (viz., (1.3.3), (1.3.4) and (1.3.6)), it is enough to define a matrix **P** of order $(v-1)\times v$ satisfying $\mathbf{P1} = \mathbf{0}$ and $\mathbf{PP}' = \mathbf{I}_{v-1}$ and identify $\boldsymbol{\theta}$ with a full set of orthonormal treatment contrasts formed by the rows of $\mathbf{P}$. It may be noted that this yields $\boldsymbol{\Sigma} = (\mathbf{PCP}')^{-1}$. We refer to Kiefer (1958) for details.

Remark 1.3.1 Kempthorne (1956) established that the efficiency factor of a block design defined in terms of the average variance of *all* paired treatment comparisons is related to $\Sigma 1/x_i$ and that the A-optimality criterion, as defined above, thus seeks to maximize the efficiency factor.

Remark 1.3.2 It may be noted that the D-optimality criterion as in (1.3.8) is *invariant* w.r.t. *any non-singular* transformation of $\boldsymbol{\theta}$ in the sense that a design which is D-optimal for $\boldsymbol{\theta}$ also remains so for $\mathbf{M}\boldsymbol{\theta}$ where $\mathbf{M}$ is *any non-singular* $(p{\times}p)$ matrix. Likewise, it may be seen that the A– and E–optimality criteria are both invariant only w.r.t. *any orthogonal* transformation of $\boldsymbol{\theta}$. With reference to a block design, for example, this would mean that a D-optimal design minimizes the generalized variance of *any* full set of $(v-1)$ linearly independent treatment contrasts while an A– (E–)optimal design minimizes the average variance (maximum eigenvalue of the dispersion matrix) of *any* orthonormal set of $(v-1)$ treatment contrasts.

Remark 1.3.3 For some *specific* inference problems involving the treatment effects, the appropriate optimality criteria may *not* be symmetric in the components of the C-matrix in the sense that $\Phi(\mathbf{C})$ may *not* equal $\Phi(\mathbf{G}_g'\mathbf{C}\mathbf{G}_g)$ for some permutation g. An example of this is that of comparing a set of *test* treatments with a *control*. This will be studied in considerable details in Chapter Seven.

Remark 1.3.4 Kiefer (1959) also introduced the notions of L– and M–optimality criteria based on considerations of behaviour of the power function of the test for H_0: $\boldsymbol{\theta} = \mathbf{0}$ over certain classes of alternatives. We omit the details.

Remark 1.3.5 There is a vast literature on *Optimal Regression Designs* (the study being initiated long time back by Smith (1918)). The underlying optimality theory relates to a search for an *approximate design.* We refer to Fedorov (1972) and Silvey (1980) for an account of such optimal designs. By contrast, the theory of optimal designs presented here will refer to *exact discrete designs* in various set-ups such as block designs, row-column designs and so on.

Remark 1.3.6 It must be noted that *all* the optimality criteria presented in this subsection are functions of the eigenvalues of the C-matrix. This relatively facilitates the study of optimality. However, there may be other criteria which are statistically meaningful and, yet, *not* exclusively functions of the eigenvalues alone. One such criterion is due to Takeuchi (1961) and is discussed in the next subsection along with two less familiar criteria suggested by Sinha (1970) and Magda (1979).

3.4 Some Further Criteria

Takeuchi (1961) argued that since in a block design we are primarily interested in *paired treatment comparisons*, an *optimal* design may seek to *minimize* the *maximum* variance of the corresponding estimates. To see that this criterion is *not* exclusively a function of the eigenvalues of the underlying C-matrix, we take the following example. Let

$$\mathbf{C}_1 = \begin{bmatrix} 7 & 0 & -7 \\ 0 & 7 & -7 \\ -7 & -7 & 14 \end{bmatrix} \text{ and } \mathbf{C}_2 = \begin{bmatrix} 13 & -9 & -4 \\ -9 & 10 & -1 \\ -4 & -1 & 5 \end{bmatrix}$$

be two competing **C**-matrices. It may be checked that they have the same eigenvalues viz., $\lambda_1 = 7$ and $\lambda_2 = 21$ but the maximum variance for a paired comparison is 2/7 for $\mathbf{C}_1$ and 13/49 for $\mathbf{C}_2$.

This criterion has been called **MV**-optimality criterion by Jacroux (1983).

Another criterion suggested by Sinha (1970) is based on the idea of *minimizing* the *distance* between the true parameter value and its estimate in a *stochastic* sense. Specifically, if $\boldsymbol{\theta}_1$ is the parameter vector of interest and $\hat{\boldsymbol{\theta}}_1$ is its estimate, we seek to minimize $Pr\{\|\hat{\boldsymbol{\theta}}_1-\boldsymbol{\theta}_1\|\geq\epsilon\}$ uniformly in $\epsilon>0$. Here $\|\hat{\boldsymbol{\theta}}_1-\boldsymbol{\theta}_1\|$ refers to Euclidean distance between $\hat{\boldsymbol{\theta}}_1$ and $\boldsymbol{\theta}_1$.

Magda (1979) introduced the notion of *Schur-optimality* via Schur-convex functions as follows. For any vector $\mathbf{x}(n\times 1)$, a real-valued function $\phi(\mathbf{x})$ satisfying $\phi(\mathbf{Sx})\leq\phi(\mathbf{x})$ for every doubly stochastic matrix $\mathbf{S}$, is said to be Schur-convex. Such a function is permutation-invariant in the sense that $\phi(\mathbf{x}) = \phi(\mathbf{Px})$ for *every* permutation matrix $\mathbf{P}$.

Let $\mathbf{x}(\mathbf{C})$ denote the vector of non-zero eigenvalues of $\mathbf{C}$. Then a *Schur-optimality* criterion seeks to minimize $\Phi(\mathbf{C}) = \phi(\mathbf{x}(\mathbf{C}))$ among all relevant **C**-matrices for a given Schur-convex function ϕ. The class of all such criteria includes **A**-, **D**-, **E**-, Φ_p- and the generalized optimality criteria and certainly much more. As for example, the Schur-convex function $\phi(\mathbf{x})$ defined as $\phi(\mathbf{x}) = \{\max(\mathbf{x})-\min(\mathbf{x})\}^{1/2}$ is *not* covered by the type of functions under generalized optimality criteria. On the other hand, **MV**-optimality criterion is *not* covered by Schur-criteria as the former is *not* exclusively a function of the eigenvalues of **C**.

In the next subsection we shall discuss the inter-relationships among the various criteria presented in subsections 3.3 and 3.4. We refer to Hedayat (1981) for an excellent review of most of the optimality criteria suggested in the literature.

3.5 Relationships Among Various Criteria

It is important to study the interrelationships among the various criteria because if a class of optimality criteria is included in a larger class, a design which is optimal w.r.t. each criterion in the larger class is optimal w.r.t. each criterion in the smaller class.

It can be verified that all the criteria considered so far satisfy (i), (ii), (iii) and (iv)′ listed in subsection 3.1 and hence are covered by extended universal optimality. We have already seen through an example in subsection 3.1 that one can construct optimality functionals which satisfy (iv)′ but not (iv) and hence are *not* covered by Kiefer's universal optimality.

It can be shown that **A**-, **D**-, **E**- and **MV**-optimality criteria are all covered by the usual notion of universal optimality i.e., these criteria satisfy Kiefer's conditions. It is *not* known if the distance criterion satisfies Kiefer's conditions.

We now examine the relation between Schur-criteria and the condition of convexity on the optimality criteria as used in the usual notion of universal optimality.

Since every stochastic matrix can be written as a convex sum of permutation matrices

$$\phi(\mathbf{Sx}) \leq \phi(\mathbf{x}) \Leftrightarrow \phi(\sum_g \alpha_g \mathbf{G}_g \mathbf{x}) \leq \phi(\mathbf{x})$$

where α_g's are non-negative with $\sum \alpha_g = 1$. Thus Schur-convexity looks somewhat like weak convexity which states that $\Phi(\sum t_g \mathbf{C}_g) \leq \Phi((\sum t_g)\mathbf{C})$ where t_g's are non-negative integers. However, a very important difference is that ϕ is defined on the set of vectors whose elements are non-zero eigenvalues of the $\mathbf{C}$-matrices whereas Φ is defined on the set of $\mathbf{C}$-matrices. To appreciate the distinction between the two we consider a particular Schur-criterion $\phi(\mathbf{x})$ and define $\Phi(\mathbf{C}) = \phi(\mathbf{x})$. We now examine if it satisfies $\Phi(\sum b_g \mathbf{C}_g) \leq \Phi((\sum b_g)\mathbf{C})$ where b_g's are non-negative integers with $1 \leq b_g < \infty$. Let $\beta_g = b_g / \sum b_g$ so that $\beta_g \geq 0$, $\sum \beta_g = 1$. A theorem due to Lidskii stated in Kato (1966) (Theorem 6.10, pp. 126) states that corresponding to any such set of β_g's, we can find a doubly stochastic matrix $\mathbf{S}(\beta)$ such that the non-zero eigenvalues of $\sum \beta_g \mathbf{C}_g$ are given by $\mathbf{S}(\beta)\mathbf{x}$. Thus, if ϕ is Schur-convex, then

$$\begin{aligned}\Phi(\sum b_g \mathbf{C}_g) &= \phi(\mathbf{x}(\sum b_g \mathbf{C}_g)) = \phi((\sum b_g)\mathbf{x}(\sum \beta_g \mathbf{C}_g)) \\ &= \phi((\sum b_g)\mathbf{S}(\beta)\mathbf{x}) \leq \phi((\sum b_g)\mathbf{x}) = \Phi((\sum b_g)\mathbf{C}).\end{aligned}$$

This demonstrates that every Schur-optimality criterion is weakly convex.

Denote by Φ_{SC} the class of optimality functionals Φ satisfying (i), (ii), (iii) of subsection 3.1 and also Schur convexity. It is then evident that $\Phi_{SC} \subseteq \Phi_{WC}$. Bondar (1983) has shown that $\Phi_{C(\lambda)} \subseteq \Phi_{SC}$ where $\Phi_{C(\lambda)}$ stands for the class of optimality criteria which are functions of the eigenvalues and satisfy (i), (ii), (iii) and (iv)'' of subsection 3.1. It thus turns out that $\Phi_{C(\lambda)} \subseteq \Phi_{SC} \subseteq \Phi_{WC} \subseteq \Phi_{S}$.

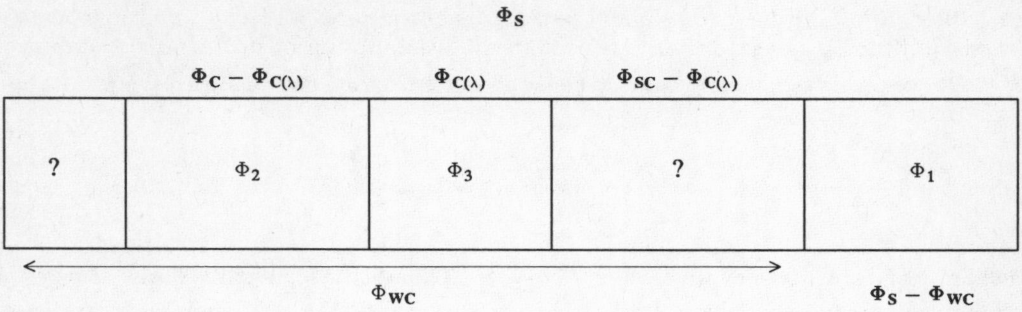

$\Phi_1 \equiv$ Distance Criterion

$\Phi_2 \equiv$ **MV**-optimality criterion

$\Phi_3 \equiv \Phi_p$-criterion

It may be remarked that in most cases the symmetry requirement (iv)′ is much easier to verify than the stronger requirement of convexity.

3.6 Concluding Remarks

It would be instructive to see the connection among majorization, Schur-convexity and ψ_f-optimality in the search for an optimal design. The key references are Marshall and Olkin (1979), Cheng (1983) and Bondar (1983). For the sake of completeness, we give below the definition of weak majorization.

A vector $\mathbf{x} = (x_1, x_2, \ldots, x_p)'$ with $0 < x_1 \leq x_2 \leq \cdots \leq x_p$ is said to be weakly upper majorized by another vector $\mathbf{y} = (y_1, y_2, \ldots, y_p)'$ with $0 < y_1 \leq y_2 \leq \cdots \leq y_p$, written symbolically as $\mathbf{x} <^w \mathbf{y}$, if

$$\sum_{i=1}^{k} x_i \geq \sum_{i=1}^{k} y_i, \quad 1 \leq k \leq p.$$

The following are equivalent definitions.

(i) $\mathbf{x} <^w \mathbf{y}$,

(ii) $\mathbf{x} = \mathbf{P}\mathbf{y}$ for some doubly superstochastic matrix $\mathbf{P}$,

(iii) $\sum_1^p f(x_i) \leq \sum_1^p f(y_i)$ for all continuous non-increasing convex functions f.

See Marshall and Olkin (1979) for the definition of a doubly superstochastic matrix. The following statements are equivalent:

(a) For a design d^*, $\mathbf{x}(\mathbf{C}_{d^*})$ is weakly upper majorized by $\mathbf{x}(\mathbf{C}_d)$ where d refers to any other competing design,

(b) d^* is optimum w.r.t. every $\Phi \epsilon \Phi_{C(\lambda)}$,

(c) d^* is optimum w.r.t. every $\Phi \epsilon \Phi_{SC}$ for which $\Phi = g(\mathbf{x}(\mathbf{C}))$ with g non-increasing (and Schur-convex),

(d) d^* is ψ_f-optimum for every (continuous) non-increasing convex f.

The equivalence (a) $\Leftrightarrow$ (d) is a restatement of (i) $\Leftrightarrow$ (iii). The equivalence (a) $\Leftrightarrow$ (b) follows from Bondar (1983). The equivalence (a) $\Leftrightarrow$ (c) follows in a routine manner by using the fact that g is non-increasing.

It may be remarked that the use of weak upper majorization (as a tool) has *not* yet been popular in the search for specific optimal designs. In Chapter Four, we will have occasion to state one result in this direction.

REFERENCES

Blackwell, D. (1951). Comparison of Experiments. *Proc. Second Berkeley Sym. Math. Statist. Prob.*, 93-102, Univ. California Press.

Blackwell, D. (1953). Equivalent comparisons of experiments. *Ann. Math. Statist.*, *24*, 265-272.

Blackwell, D. and Girshick, M.A. (1954). *Theory of games and statistical decisions.* Wiley, New York.

Bondar, J.V. (1983). Universal optimality of experimental designs: definitions and a criterion. *Canadian Jour. Statist.*, *11*, 325-331.

Chakrabarti, M.C. (1962). *Mathematics of Design and Analysis of Experiments.* Asia Publishing House, Bombay.

Cheng, C.S. (1978). Optimality of certain asymmetrical experimental designs. *Ann. Statist.*, *6*, 1239-1261.

Cheng, C.S. (1985). Commentary on papers [22], [55], [60], [61] in *Jack Carl Kiefer Collected Papers: III Design of Experiments.* Springer-Verlag, 695-700.

Eccleston, J.A. and Hedayat, A.S. (1974). On the theory of connected designs: Characterization and optimality. *Ann. Statist.*, *2*, 1238-1255.

Fedorov, V.V. (1972). *Theory of optimal experiments.* Academic Press, New York.

Hedayat, A.S. (1981). Study of optimality criteria in design of experiments. In *Statistics and Related Topics.* Ed. M. Csorgo *et al.* North-Holland.

Jacroux, M. (1983). Some minimum variance block designs for estimating treatment differences. *J.R. Statist. Soc. B*, *45*, 70-76.

Kato, T. (1966). *Perturbation Theory for Linear Operators.* A Series of Comprehensive Studies in Mathematics. Springer-Verlag.

Kempthorne, O. (1956). The efficiency factor of an incomplete block design. *Ann. Math. Statist.*, *27*, 846-849.

Kiefer, J. (1959). Optimum experimental designs. *J.R. Statist. Soc. B, 21*, 272-319.

Kiefer, J. (1975). Construction and optimality of generalized Youden designs. In *A Survey of Statistical Design and Linear Models.* (J.N. Srivastava, ed.). Amsterdam: North-Holland Publishing Co., (1975) 333-353.

Magda, C.G. (1979). On E-optimal block designs and Schur optimality. *Ph.D. Thesis.* University of Illinois at Chicago.

Marshall, A.W. and Olkin, I. (1979). *Inequalities: Theory of Majorizaton and Its Applications.* Academic Press.

Shah, K.R. (1960). Optimality criteria for incomplete block designs. *Ann. Math. Statist., 31*, 791-794.

Shah, K.R. and Khatri, C.G. (1973). Connectedness in row-column designs. *Comm. in Statist., 2*, 571-573.

Shrikhande, S.S. (1951). Designs for two-way elimination of heterogeneity. *Ann. Math. Statist., 22*, 235-247.

Silvey, S.D. (1980). *Optimal design.* Chapman and Hall, London.

Sinha, B.K. (1970). On the optimality of some designs. *Cal. Statist. Assoc. Bull., 19*, 1-22.

Smith, K. (1918). On the standard deviations of adjusted and interpolated values of an observed polynomial function and its constants and the guidance they give towards a proper choice of the distribution of observations. *Biometrika, 12*, 1-85.

Takeuchi, K. (1961). On the optimality of certain type of PBIB designs. *Rep. Statist. Appl. Res. Un. Japan Sci. Engrs., 8*, 140-145.

Wald, A. (1943). On the efficient design of statistical investigations. *Ann. Math. Statist., 14*, 134-140.

Yeh, C.M. (1986). Conditions for universal optimality of block designs. *Biometrika, 73*, 701-706.

Yeh, C.M. (1988). A class of universally optimal binary block designs. *J. Statist. Planning and Inference, 18*, 355-361.

CHAPTER TWO

BLOCK DESIGNS: GENERAL OPTIMALITY

1. Introduction

In this Chapter we intend to present the *essential* results known so far regarding optimality of certain classes of block designs w.r.t. some general optimality criteria. Unless otherwise stated, the competing class of designs, to be denoted by $\mathbf{D}(b,v,k)$, will comprise of *all* connected designs in which v treatments are compared in b blocks each of size k. While presenting the results on optimal block designs we will deliberately restrict ourselves to the experimental set-up of blocks of equal sizes. With blocks of unequal sizes, the assumption of homogeneity of error variances is itself questionable and, further, the optimality results are seen to depend on too many design parameters. In Chapter Three (Section 7), we have mentioned some current work on optimality results involving the latter set-up.

We have organized the presentation of the material in the following order. In section 2, we give a formal definition of *Balanced Block Designs* (**BBD**s) and establish their universal optimality in the extended sense developed in subsection 3.2 of Chapter One. In view of the discussion in subsection 3.5 of Chapter One, it follows that the **BBDs** are optimal w.r.t. *any* generalized criteria of Cheng (1978) as also w.r.t. the criteria of Takeuchi (1961) and of Sinha (1970). It may be mentioned that no designs, other than the **BBDs**, are known to be universally optimal in the entire class $\mathbf{D}(b,v,k)$. However, Yeh (1986, 1988) produced a series of designs which are universally optimal in a restricted class.

Since the **BBDs** may *not* be available for many combinations of values of (b,v,k), the search for designs optimal w.r.t. a less general class of criteria and/or within a smaller class of competing designs becomes relevant. In section 3, we essentially deal with the generalized optimality criteria and establish optimality of certain classes of designs w.r.t. these criteria. Such optimal designs are mainly (i) some classes of the *Group Divisible Designs* (**GDDs**) and (ii) the minimal covering designs for $6t+5$ treatments in blocks of size 3. Further, duals of *equireplicate optimal* designs are also shown to be *optimal* within the *restricted* class of equireplicate designs.

2. Universal Optimality of the BBDs

We start with the following formal definition of a **BBD** as given by Kiefer (1958).

For given b, v and k, a design with incidence matrix $\mathbf{N} = ((n_{ij}))$ is said to be a *Balanced Block Design* (**BBD**) if the elements of $\mathbf{N}$ *satisfy the following:*

(i) $n_{ij} = [k/v]$ or $[k/v] + 1$ where $[x] =$ largest integer *not* exceeding x;

(ii) $r_i = \sum_{j=1}^{b} n_{ij}$ is the same for all $1 \leq i \leq v$;

(iii) $\lambda_{ii'} = \sum_{j=1}^{b} n_{ij} n_{i'j}$ is the same for all $1 \leq i \neq i' \leq v$.

Clearly, (i) and (ii) imply

(iv) $\lambda_{ii} = \sum_{j=1}^{b} n_{ij}^2$ is the same for all $1 \leq i \leq v$.

A **BBD** reduces to a *Balanced Incomplete Block Design* (**BIBD**) when $k<v$. Also it reduces to a *Randomized Block Design* (**RBD**) when $k = v$. It may be noted that the incidence matrix **N** of a **BBD** has the representation

$$\mathbf{N}(v \times b) = [k/v]\mathbf{J}_{v \times b} + \mathbf{N}^*(v \times b)$$

where $\mathbf{J}_{p \times q} = \mathbf{11}'$ is a matrix of order $p \times q$ with all elements unity, and $\mathbf{N}^*(v \times b)$ is the incidence matrix of a **BIBD**. Further, among other things, it is necessary that bk/v is an integer which we will denote by r.

The above representation of **N** essentially means that a **BBD** with *parameters* (b,v,k) is to be regarded as a combination of $[k/v]$ copies of a Randomized (Complete) Block Design (**RBD**) with parameters (b,v) and a **BIBD** with parameters $(b,v,k^* = k-v[k/v])$. In view of this, existence of a **BBD** is equivalent to the existence of an underlying **BIBD** and, as such, no new problem of construction arises for the **BBDs**. It is clear that Fisher's inequality $(b \geq v)$ which holds for a **BIBD**, also holds for *any* **BBD** resulting into that **BIBD**.

We now establish extended universal optimality of the **BBDs**. In view of *Proposition 1 (Restated)* in subsection 3.2 of Chapter One, it is enough to verify that the **C**-matrix of a **BBD** with parameters (b,v,k) is completely symmetric (c.s.) and it maximizes $tr(\mathbf{C})$ among *all* designs in $\mathbf{D}(b,v,k)$. Denote by d^* the **BBD** and by $\mathbf{C}^*$ its **C**-matrix. From (1.2.4) it follows that for any design in $\mathbf{D}(b,v,k)$,

$$\left.\begin{aligned} C_{ii} &= \sum_j n_{ij} - \sum_j n_{ij}^2/k = r_i - \lambda_{ii}/k \\ \text{and } C_{ii'} &= -\sum_j n_{ij}n_{i'j}/k = -\lambda_{ii'}/k \end{aligned}\right\} \tag{2.2.1}$$

From the definition of the **BBD**, it follows that $\mathbf{C}^*$ has diagonal elements *all* equal and also off-diagonal elements *all* equal. Hence it is *completely symmetric* (c.s.).

To show that $\mathbf{C}^*$ has maximum trace, we note that for an arbitrary design,

$$tr(\mathbf{C}) = bk - k^{-1}\sum_i \sum_j n_{ij}^2.$$

Clearly, n_{ij}'s satisfy $\sum_i n_{ij} = k$ for all $1 \leq j \leq b$. Hence, rewriting $\sum_i \sum_j n_{ij}^2$ as $\sum_j (\sum_i n_{ij}^2)$, it can be seen that $tr(\mathbf{C})$ is maximized when $n_{ij} = [k/v]$ or $[k/v]+1$ for all $1 \leq i \leq v$ and for all $1 \leq j \leq b$. Clearly, this is the case with the **BBD**. Thus a **BBD** is seen to be universally optimal among *all* connected designs. It is evident that a **BBD** is **A**-, **D**- and **E**-optimal as well as optimal w.r.t. any generalized optimality criterion.

Remark 2.2.1. It must be noted that trace maximization takes place iff n_{ij}'s assume only the values $([k/v], [k/v]+1)$ and, as a matter of fact, it is easy to construct plenty of designs having this feature. Such designs are called *binary* or *generalized binary* according as $k<v$ or $k\geq v$. On the other hand, very few designs yield completely symmetric **C**-matrices. It can be verified that *only* the **BBD**s attain *both* the features *except* when $v=2$ and k is odd. In this exceptional case, *any* design has a completely symmetric **C**-matrix and there are designs with *unequal* number of replications which maximize the trace. We mention in passing that the balanced *ternary* designs given by Tocher (1952) have completely symmetric **C**-matrices. However, such designs do *not* provide maximum trace since the n_{ij}'s take three distinct values.

It may be of interest to note that Yeh (1986, 1988) came out with a series of designs whose **C**-matrices are *not* completely symmetric. However, such designs turn out to be universally optimal in the class of generalized binary designs. The design parameters are b, v, $k = \pm 1 \pmod{v}$. It turns out that for such parameter values, *any* generalized binary design which is equireplicate or nearly so, is universally optimal. Also see Gaffke (1981) in this context.

3. Optimality of Some Classes of Asymmetrical Designs w.r.t. the Generalized Criteria

3.1 Motivation and Summary

As we have seen earlier, universally optimal designs are characterized by the property of trace maximization and complete symmetry of the **C**-matrix which are achieved largely by the **BBD**s. The desirability of trace maximization can be interpreted in the following way. In the first place, as the **C**-matrix plays the role of the information matrix for the treatment effects parameters, one would like to prefer a design with maximum information content at least in the sense of maximum possible value of the trace of the information matrix. Secondly, when $k<v$, trace maximization takes place iff the design is binary. In such a case, there are precisely $b(k-1)$ within block treatment comparisons available whereas for a non-binary design, the number of such comparisons would be less. This leads to the strong feeling that an optimal design w.r.t. *any symmetric criterion* is highly likely to be binary whenever $k<v$, and, otherwise, *generalized* binary in the sense that $n_{ij} = [k/v]$ or $[k/v]+1$, for all $1\leq i\leq v$, $1\leq j\leq b$. In fact, there is *no* known example where the best binary or generalized binary design is *not* optimal. In the particular case of $k=2$, it is trivially seen that a block with identical treatments does *not* contribute to treatment comparisons and, hence, can be replaced by a block with any two distinct treatments. This rules out designs with block(s) having repeated treatments at least when $k=2$.

The other important feature of the **BBD**s is that they possess completely symmetric **C**-matrices. Denoting the non-zero eigenvalues of a **C**-matrix by $x_1, x_2, \ldots, x_{v-1}$, it is easily seen that for a **BBD**, x_i's are all equal and, hence, $\Sigma(x_i-\bar{x})^2 = 0$ where $\bar{x} = \Sigma x_i/(v-1)$.

It may be noted that for *any* binary or generalized binary design, $\bar{x} = \dfrac{b(k-1)}{v-1} - \dfrac{b}{k(v-1)}[k/v]\{2k-v-v[k/v]\}$. Therefore, it is true that a **BBD** also minimizes Σx_i^2 among *all* binary/generalized binary designs. Thus it seems reasonable that if we wish to work with designs having maximum trace, a design with the least value of Σx_i^2 is likely to be a *reasonably good* choice. We may further justify this w.r.t.

the Generalized Optimality Criteria (Vide (1.3.1) and (1.3.2), Chapter One). Considering a *formal* expansion of $f(x_i)$ about $\bar{x}$, we observe that

$$\Sigma f(x_i) \approxeq (v-1)f(\bar{x}) + \frac{1}{2}\Sigma(x_i-\bar{x})^2 f''(\bar{x}).$$

Hence, minimization of $\Sigma(x_i-\bar{x})^2$ hopefully leads to the least value of $\Sigma f(x_i)$ as $f''(\bar{x})>0$, $f(.)$ being convex.

The above discussion justifies the use of **S**– and **(M,S)**–optimality criteria given in Chapter One (immediately after (1.3.4)). These two criteria do *not* as such have any statistical interpretation. However, they may be regarded as tools to get to the designs with generalized optimality. Indeed, they have been found extremely useful in this venture. Theoretical developments based on the above were initiated by Conniffe and Stone (1974,'75) who also succeeded in establishing **A**–optimality of some classes of **GDDs**. Cheng (1978) made an ingeneous use of Conniffe and Stone's technique to deduce a powerful result which has been found extremely useful in settling generalized optimality of the **GDDs** studied by Conniffe and Stone (1975). Cheng's result has also been used in many other contexts. In particular, Roy and Shah (1984) used it in deducing generalized optimality of some classes of *minimal covering* designs in cases where bk is *not* divisible by v. Further, Cheng (1979) categorically derived forms of optimal incomplete block designs for four varieties.

Duals of *equireplicate* designs with the property of generalized optimality in the class $\mathbf{D}(b,v,k)$ are also seen to possess generalized optimality property within the *restricted* class of equireplicate designs with parameters $\tilde{b} = v$, $\tilde{v} = b$ and $\tilde{k} = bk/v$. This completes an overview of the essential results known so far w.r.t. the generalized optimality criteria.

3.2 A Theorem for Generalized Optimality

The objective is to obtain a design d^* in the design class $\mathbf{D}(b,v,k)$ which minimizes $\Sigma f(x_i)$ for *every* optimality functional f satisfying (1.3.2). This will lead to d^* being optimal w.r.t. any generalized optimality criterion.

The minimization is carried out in two stages following Conniffe and Stone (1974). The precise results are given below.

Lemma 2.3.1. For fixed $A = \sum_1^{v-1} x_i$ and $B = \sum_1^{v-1} x_i^2$ subject to $A^2>B>A^2/(v-1)$, for *any* function f satisfying (1.3.2), $\psi_f = \sum_1^{v-1} f(x_i)$ attains its minimum when exactly one of the x_i's assumes the value $\{A+\delta(v-2)P\}/(v-1)$ and the rest are each equal to $\{A-\delta P\}/(v-1)$ where $P^2 = B-A^2/(v-1)$ and $\delta = \sqrt{(v-1)/(v-2)}$.

Proof. Let $S(A,B) = \{(x_1 x_2, \ldots, x_{v-1}) | x_i>0$ for all i, $\Sigma x_i = A$ and $\Sigma x_i^2 = B\}$. Since $f(0) = \lim_{x\to 0+} f(x) = +\infty$, it follows that $\psi_f = \sum_1^{v-1} f(x_i)$ attains its minimum at an interior point of $S(A,B)$. We write, then, $\sum_1^{v-1} f(x_i) = \sum_1^{v-2} f(x_i) + f(A-\sum_1^{v-2} x_i)$ and consider the problem of minimizing it subject to $\sum_1^{v-2} x_i^2 + (A-\sum_1^{v-2} x_i)^2 = B$, $x_i>0$, $1\le i\le v-2$, $\sum_1^{v-2} x_i<A$.

Differentiating

$$\sum_1^{\nu-2} f(x_i) + f(A-\sum_1^{\nu-2} x_i) + \theta\{B-\sum_1^{\nu-2} x_i^2 - (A-\sum_1^{\nu-2} x_i)^2\}$$

w.r.t. each x_i and equating the derivative to zero, one gets

$$f'(x_i) - f'(x_{\nu-1}) + \theta(-2x_i + 2x_{\nu-1}) = 0, \quad 1 \leq i \leq \nu-2$$

which means that either $x_i = x_{\nu-1}$ or if $x_i \neq x_{\nu-1}$,

$$2\theta = \frac{f'(x_i)-f'(x_{\nu-1})}{x_i - x_{\nu-1}}, \quad 1 \leq i \leq \nu-2.$$

For fixed $x_{\nu-1}$, the strict concavity of f' implies that the above equation has a unique solution $x_i \in (0,A)$. Therefore, over $S(A,B)$, all the stationary points of ψ_f can have exactly two distinct co-ordinates. Call them μ and μ' with $\mu > \mu'$. If μ has multiplicity n, $1 \leq n \leq \nu-2$, we set

$$n\mu + (\nu-1-n)\mu' = A, \; n\mu^2 + (\nu-1-n)\mu'^2 = B$$

and $\mu > \mu'$ and solve readily for μ and μ' as

$$\left.\begin{aligned} \mu(n;A,B) &= \{A + \sqrt{(\nu-1)(\nu-1-n)/n}P\}/(\nu-1) \\ \mu'(n;A,B) &= \{A - \sqrt{n(\nu-1)/(\nu-1-n)}P\}/(\nu-1) \end{aligned}\right\} \qquad (2.3.1)$$

Here $P^2 = B - \dfrac{A^2}{\nu-1}$.

But this yields $\psi_f = nf(\mu(n;A,B)) + (\nu-1-n)f(\mu'(n;A,B)) = \psi_f(n)$ (say) and since f' is strictly concave, one can readily verify that $\psi_f(n)\uparrow$ in n. Hence, we take $n = 1$ so that, finally, $\min\limits_{S(A,B)} \psi_f = f(\mu(1;A,B)) + (\nu-2)f(\mu'(1;A,B)) = M_f(A,B)$ (say) where μ and μ' are as in (2.3.1). We have explicitly

$$\begin{aligned} M_f(A,B) = {} & f(\{A + \sqrt{(\nu-1)(\nu-2)}P\}/(\nu-1)) \\ & + (\nu-2)f(\{A - \sqrt{(\nu-1)/(\nu-2)}P\}/(\nu-1)) \end{aligned} \qquad (2.3.2)$$

This completes the proof of the Lemma.

Remark 2.3.1. For $f(x) = 1/x$, this result was first established by Conniffe and Stone (1974). The above proof is due to Cheng (1978). Recently, Cheng (1987) provided a solution to a more general optimization problem of the form:

Minimize $\sum f(x_i)$

Subject to $\sum x_i = A$, $\sum g(x_i) = B$, $x_i \geq 0$.

Lemma 2.3.1 presented above is a special case of this. Cheng (1987) also provided a set of sufficient conditions for a design to minimize $\sum f(x_i)$. His Theorem 3.1 and Theorem 3.2 are generalizations of Theorem 2.3.2 stated below. We mention in passing that Theorem 2.3.2 is quite adequate for deriving optimality results presently available.

Remark 2.3.2. The above Lemma has indeed provided a new direction and re-thinking in the search for optimal designs in situations where the **BBDs** do *not* exist. The expression (2.3.2) clearly suggests that a design with the **C**-matrix having only *two* distinct eigenvalues and, moreover, the larger one having multiplicity unity, is likely to be a promising candidate.

The following result (due to Cheng (1978)) explicitly provides further conditions on A and P which ensure optimality of designs having such **C**-matrices. We may note in this context that $A = tr(\mathbf{C})$ and $B = tr(\mathbf{C}^2) = \sum_i \sum_j c_{ij}^2$.

Theorem 2.3.2. Suppose there exists a design d^* in $\mathbf{D}(b,v,k)$ such that its **C**–matrix has roots

$$\left.\begin{array}{ll} \mu^* = \{A^*+\delta(v-2)P^*\}/(v-1) & \text{with multiplicity } 1 \\ \mu'^* = \{A^*-\delta P^*\}/(v-1) & \text{with multiplicity } (v-2) \end{array}\right\} \tag{2.3.3}$$

where A^* and P^* are the corresponding values of A and P for the design d^* and $\delta = \sqrt{(v-1)/(v-2)}$.

Suppose further that

$$\left.\begin{array}{l} \text{(i) } A^* \geq A \\ \text{(ii) } A^*-\delta P^* \geq A-\delta P \end{array}\right\} \tag{2.3.4}$$

for *all* pairs (A,P) based on competing designs in $\mathbf{D}(b,v,k)$.

Then d^* is optimal w.r.t. *every* generalized optimality criterion.

Proof. At the outset, we note that for any competing design, $\psi_f \geq M_f(A,B)$. Next, we recall that f is monotone decreasing and convex. It is then easy to verify that the function $M_f(A,B)$ in (2.3.2) is a decreasing function of A for fixed P and, further, it is also an increasing function of P for fixed A. If now $P^* \leq P$, then obviously $M_f(A,B) \geq M_f(A,B^*) \geq M_f(A^*,B^*)$. On the other hand, if $P^* \geq P$, then $(v-1)\mu^* = A^* + \delta(v-2)P^* \geq A + \delta(v-2)P = \mu(v-1)$. Further, $(v-1)\mu'^* = A^*-\delta P^* \geq A-\delta P = \mu'(v-1)$ by the hypothesis of the Theorem. Hence, because of the monotone decreasing nature of f, it follows that $M_f(A,B) \geq M_f(A^*,B^*)$.

This establishes the claim.

Remark 2.3.3. We note that the (sufficient) condition (i) in (2.3.4) relates to trace maximization which can be easily verified. However, the condition (ii) in (2.3.4) does *not* have any obvious interpretation and it is relatively much harder to verify. The above theorem certainly holds if the condition (ii) in (2.3.4) is replaced by $B^* < B$ or $P^* < P$ but these may prove false or, else, even harder to verify.

Remark 2.3.4. If the class of competing designs is *restricted* to generalized binary designs only, then the condition (ii) in (2.3.4) is equivalent to **S**–optimality or (**M**,**S**)–optimality.

3.3 Optimality of Some Classes of GDDs

We start by observing that the condition (i) in (2.3.4) simply means that we should look for a generalized binary design. Let us now examine the implication of (2.3.3) as regards the structure of the **C**–matrix. We have

$$\mathbf{C} = \mathbf{r}^{\delta} - k^{-1}\mathbf{N}\mathbf{N}'$$
$$= \mu^* \xi\xi' + \mu'^*(\mathbf{I}_v - \mathbf{J}/v - \xi\xi') \quad \text{(by (2.3.3))}$$

This gives $C_{ii} = r_i - k^{-1}\sum_j n_{ij}^2 = (\mu^* - \mu'^*)\xi_i^2 + \mu'^*(1 - 1/v)$ and

$C_{ii'} = -k^{-1}\sum_j n_{ij}n_{i'j} = (\mu^* - \mu'^*)\xi_i\xi_{i'} - \mu'^*/v.$

If the design is *generalized binary* and *equireplicate*, then $r_i - k^{-1}\sum_j n_{ij}^2$ is the same for *all* i and, consequently, ξ_i^2 is the same for *all* i $(= \frac{1}{v}$, as $\sum_1^v \xi_i^2 = 1)$. Since $\sum_1^v \xi_i = 0$, this means that v must be an even integer and that $\xi_i = 1/\sqrt{v}$ in half of the cases, and $\xi_i = -1/\sqrt{v}$ in the other half. Identifying the two halves as forming two groups each consisting of $v/2 = n$ (say) treatments, we can immediately see from the above expressions of the $C_{ii'}$'s that the underlying design conforms to a generalized or ordinary **GDD** depending on whether $k >$ or $< v$.

For completeness, we give below a formal definition of a **GDD**.

Definition 2.3.1. A design in $\mathbf{D}(b,v,k)$ with incidence matrix $\mathbf{N}$ is said to be a **GDD** or a generalized **GDD** if

(i) $n_{ij} = [k/v]$ or $[k/v]+1$ for all $1 \le i \le v$, $1 \le j \le b$

(ii) r_i's are all equal

(iii) the treatments can be divided into m groups of n each such that $\lambda_{ii'} = \lambda_1$ if i and i' are in the same group, and $\lambda_{ii'} = \lambda_2$ otherwise.

For $k<v$, it gives the familiar definition of a **GDD**. When $\lambda_2 = \lambda_1 \pm 1$, a **GDD** is referred to as a *Most Balanced* (**MB**) **GDD**.

It is well-known that the **C**–matrix of a **GDD** with parameters $(b,v,k,m,n,\lambda_1,\lambda_2)$ has the eigenvalues $r - \lambda_1$ and $rk - v\lambda_2$ with multiplicities $(m-1)$ and $(v-m)$ respectively when $k<v$. In the case when $k>v$, it can be seen that the eigenvalues are

$r - \dfrac{k^*\lambda_1^*}{k}$ with multiplicity $(m-1)$

$r - \dfrac{1}{k}\{vk^*\lambda_2^* - k^*r^*(k^*-1)\}$ with multiplicity $(v-m)$

where $k = v[k/v]+k^*$, $r = b[k/v] + r^*$, $\lambda_1^* = \lambda_1 - [k/v](b[k/v]+2r^*)$,

$\lambda_2^* = \lambda_2 - [k/v](b[k/v]+2r^*)(= \lambda_2+(\lambda_1^*-\lambda_1))$. The condition (2.3.3) demands that for $m = 2$, $r - \dfrac{k^*\lambda_1^*}{k} > r - \dfrac{1}{k}[vk^*\lambda_2^* - k^*r^*(k^*-1)]$ which simplifies to $\lambda_1^* < \lambda_2^*$ or equivalently, $\lambda_1 < \lambda_2$.

Looking at condition (ii) in (2.3.4), we see that in order to maximize $A^* - \delta P^*$ for a *generalized binary design*, we may seek to minimize P^* or, equivalently, B^*. And for a **GDD**, this is achieved when $|\lambda_2 - \lambda_1| = 1$. Thus a **GDD** with $m = 2$ and $\lambda_2 = \lambda_1+1$ achieves both (2.3.3) and (2.3.4) (ii) among *all* generalized binary designs.

The above analysis shows that a **GDD** with $m = 2$ and $\lambda_2 = \lambda_1+1$ is optimal among *all generalized binary designs*. For the **A**–optimality criterion ($f(x) = \dfrac{1}{x}$), this was established by Conniffe and Stone (1975). Cheng (1978) proceeded further to verify that such a **GDD** indeed satisfies the condition (ii) in (2.3.4) for *all* competing designs in the class $\mathbf{D}(b,v,k)$. Since the generalized binary designs are *not* yet known to be superior to non-generalized binary ones, this sort of verification was necessary for extending the results to the *entire* class $\mathbf{D}(b,v,k)$.

Below we present the verification for $A^* - \delta P^* > A - \delta P$ where (A,P) correspond to a non-generalized binary design. Towards this, we first note that

$$\left.\begin{array}{rc} \text{(i)} & v = 2n \\ \text{(ii)} & kA^* = v(v-1)\lambda_1 + v(n-1) \\ \text{(iii)} & k^2P^{*2} = vn(n-1)/(v-1) \\ \text{(iv)} & \delta = \{(v-1)/(v-2)\}^{1/2} \\ \text{(v)} & k\delta P^* = n \end{array}\right\} \qquad (2.3.5)$$

Since $k\mathbf{C}$ is a matrix with integer elements, we may write

$$kA = kA^* - \alpha, \text{ for some integer } \alpha. \qquad (2.3.6)$$

Now if $\alpha \geq n$, $k(A^*-A) = \alpha \geq n = k\delta P^*$ (by (2.3.5)(v)) and, consequently, $A^* - \delta P^* \geq A - \delta P$. So we need only consider the case when $\alpha < n$. Here we will establish that

$$\delta kP \geq \delta kP^* + (A - A^*)k = n - \alpha \text{ (by (2.3.5)(v) and (2.3.6))}.$$

For this, we observe that $k^2B = tr(k^2\mathbf{C}^2) = \sum_i k^2C_{ii}^2 + \sum_{i \neq}\sum_{i'} k^2C_{ii'}^2$ where $\sum_i kC_{ii} = kA$ and $\sum_{i \neq}\sum_{i'} kC_{ii'} = -kA$. The first term $\sum_i k^2C_{ii}^2$ is *not* less than k^2A^2/v. Further the second term is bounded below by

$$[\frac{kA}{v(v-1)}]^2 h + ([\frac{kA}{v(v-1)}] + 1)^2\{v(v-1)-h\}$$

where

$$h = v(v-1)([\frac{kA}{v(v-1)}] + 1)-kA.$$

Now note that as $\alpha<n$ and $kA = kA^*-\alpha = v(v-1)\lambda_1+v(n-1)-\alpha$ (by (2.3.5)(ii)), $[\frac{kA}{v(v-1)}] = \lambda_1$ and, hence, $h = v(v-1)(\lambda_1+1)-kA = vn+\alpha$. This yields a lower bound to B from which a lower bound to P^2 is derived as

$$k^2P^2 \geq \lambda_1^2(vn+\alpha) + (\lambda_1+1)^2(v(n-1)-\alpha) - \{v(v-1)\lambda_1+v(n-1)-\alpha\}^2/v(v-1)$$

$$= (v(n-1)-\alpha)(nv+\alpha)/v(v-1) \text{ (on simplification).}$$

Finally, recalling that $\delta^2 = (v-1)/(v-2)$ and $v = 2n$, we get that $\delta^2k^2P^2 \geq (v(n-1)-\alpha)(nv+\alpha)/v(v-2)$ and this latter quantity is easily seen to be $\geq (n-\alpha)^2$ whenever $v\geq 4$, using $\alpha<n = v/2$. This is what was to be established. Thus we have succeeded in showing that certain families of generalized **GDDs** are optimal in the *entire* class w.r.t. the generalized optimality criteria.

Further to this, it may be added that such **GDDs** are *uniquely* optimal in the sense that in the presence of these designs, no other design can be optimal *even* w.r.t. a specific criterion of the form (1.3.1) satisfying (1.3.2). This, however, leaves *open* the possibility of existence of designs (other than these **GDDs**) which might also be optimal w.r.t. the point-wise limits of such criteria such as **E**–optimality.

3.4 **Optimality of a Class of Minimal Covering Designs**

Roy and Shah (1984) established optimality of a class of *minimal covering designs*. A minimal covering design for v *treatments in blocks of size* k *is formed by taking the smallest number of blocks for which* $\sum_j n_{ij}n_{i'j} \geq 1$ *for all* $i \neq i'$ can be achieved. The case when $k = 3$ has been extensively studied in the combinatorial literature. Fort and Hedlund (1958) discussed the problem of the minimal covering of pairs by triplets (i.e., $k = 3$). They first noted that for v treatments, the number of blocks required must be at least $\phi(v) = \lceil (v/3)\lceil (v-1)/2\rceil\rceil$ where $\lceil x\rceil$ denotes the smallest integer greater than or equal to x. They then showed that for all $v\geq 3$, designs with $\phi(v)$ triplets can, in fact, be constructed.

It is well-known that for $v = 6t+1$ or $6t+3$, the *Steiner Triplet Systems* (**STSs**) give minimal covering designs with $\lambda_{ii'} = 1$ for all $i \neq i'$. Clearly the **STSs** are identified as **BIBDs** and, hence, these are universally optimal. For any other value of v, a minimal covering design does *not* achieve $\lambda_{ii'} = 1$ for *all* $i \neq i'$.

Roy and Shah (1984) studied the case of $v = 6t+5$. In this case, a minimal covering design consists of $b = 6t^2+9t+4$ blocks. Methods for constructing such designs have been given in Fort and Hedlund (1958) as also in Roy and Shah (1984). The matrix $\mathbf{NN'}$ for such a design can be expressed as

$$\mathbf{NN'} = \left[\begin{array}{cc|c} 3t+3 & 3 & \mathbf{J}_{2,6t+3} \\ 3 & 3t+3 & \\ \hline \multicolumn{2}{c|}{\mathbf{J}_{6t+3,2}} & (3t+1)\mathbf{I}_{6t+3} + \mathbf{J}_{6t+3,6t+3} \end{array}\right]$$

This shows that for a minimal covering design with parameters $b = 6t^2+9t+4$, $v = 6t+5$, $k = 3$, the replication numbers are

$$r_1 = r_2 = 3t+3, \qquad r_3 = \cdots = r_v = 3t+2$$

and the $\lambda_{ii'}$'s assume only two distinct values viz.,

$$\lambda_{12} = 3, \ \lambda_{ii'} = 1 \ \text{ for } all \ (ii') \neq (1\,2).$$

It may be seen that the eigenvalues of the resulting $\mathbf{C}$–matrix are $(v+4)/3$ with multiplicity 1 and $v/3$ with multiplicity $(v-2)$. Roy and Shah (1984) verified the conditions (2.3.3) and (2.3.4) of Theorem 2.3.2 in the entire class of designs $\mathbf{D}(b,v,k)$. We omit the details.

Thus generalized optimality of such *asymmetrical unequally replicated* designs has been established.

3.5 Optimal Designs with Four Varieties

Cheng (1979) extensively studied the optimality aspects of *incomplete* block designs involving four varieties and came up with designs which are intuitively acceptable. Two cases of relevance are

$$\text{(I)} \quad k = 2, \ v = 4, \ b = 6t+s, \quad 1 \leq s \leq 5$$

$$\text{(II)} \quad k = 3, \ v = 4, \ b = 4t+s, \quad 1 \leq s \leq 3.$$

In case (I), an *optimal* design is expected to involve t copies of the **BIBD** (6,4,2) followed by s additional blocks each of size 2. These s blocks of an optimal design are found out to be as shown below.

		s		
(1)	(2)	(3a) or (3b)	(4)	(5)
(1 2)	(1 2)	(1 2) (1 2)	(1 3)	(1 3)
	(3 4)	(1 3) (3 4)	(1 4)	(1 4)
		(1 4) (1 3)	(2 3)	(2 3)
			(2 4)	(2 4)
				(3 4)

In case (II), an *optimal* design is likewise expected to consist of t copies of the **BIBD** (6,4,3) followed by s additional blocks each of size 3. These s blocks are shown below.

	s	
(1)	(2)	(3)
(1 2 3)	(1 2 3)	(1 2 4)
	(1 2 4)	(1 3 4)
		(2 3 4)

Except for the case when $b = 6t+3$, $k = 2$, these designs have, in fact, been shown to be optimal w.r.t. any generalized optimality criterion. In this exceptional case, a design with additional blocks of the form (3a) is **E**–optimal while one with those of the form (3b) is **A**– and **D**–optimal. We refer to Cheng (1979) for the proofs.

3.6 Optimality of Dual Designs

Two designs d and $\tilde{d}$ with incidence matrices **N** and $\tilde{\mathbf{N}}$ respectively are said to be *duals* of each other when **N** and $\tilde{\mathbf{N}}$ are transposes of each other. It is known that $\tilde{d}$ is connected iff d is connected.

For given b, v and k, let d be a design in the entire class $\mathbf{D}(b,v,k)$ of connected designs. Then the dual of d will have blocks of equal size iff d is equireplicate. We thus assume $bk/v = r$ (an integer) and confine our attention to the *restricted* class $\mathbf{D}^*(b,v,k,r)$ of connected *equireplicate* designs. Then the dual $\tilde{d}$ of d belongs to the class $\tilde{\mathbf{D}}^*(\tilde{b}=v, \tilde{v}=b, \tilde{k}=r, \tilde{r}=k)$ of connected *equireplicate* designs. In view of the fact that $\mathbf{N} = \tilde{\mathbf{N}}'$, we obtain

$$\left.\begin{aligned} \mathbf{C} &= r\mathbf{I}_v - k^{-1}\mathbf{N}\mathbf{N}', \\ \tilde{\mathbf{C}} &= \tilde{r}\mathbf{I}_{\tilde{v}} - \tilde{k}^{-1}\tilde{\mathbf{N}}\tilde{\mathbf{N}}' = k\mathbf{I}_b - r^{-1}\mathbf{N}'\mathbf{N} \end{aligned}\right\} \qquad (2.3.7)$$

As noted by Roy (1958), this immediately establishes a connection between the non-zero eigen-values of **C** and $\tilde{\mathbf{C}}$ in general terms. Denoting the eigen-values of **C** by $x_1, x_2, \ldots, x_{v-1}$ and those of $\tilde{\mathbf{C}}$ by $\tilde{x}_1, \tilde{x}_2, \ldots, \tilde{x}_{b-1}$, it is easily seen that

$$\left.\begin{aligned} \text{for } v<b,\quad \tilde{x}_i &= \frac{kx_i}{r}, & 1\le i\le v-1 \\ &= k, & v\le i\le b-1 \\ \\ \text{for } v>b,\quad x_i &= \frac{r}{k}\tilde{x}_i, & 1\le i\le b-1 \\ &= r, & b\le i\le v-1 \end{aligned}\right\} \tag{2.3.8}$$

Thus in the set-up of $\mathbf{D}(b,v,k,r)$, a design d_1 is *better* than another design d_2 w.r.t. a specific optimality criterion viz., minimization of $\Sigma f(x_i)$ iff $\tilde{d}_1$ is *better* than $\tilde{d}_2$ w.r.t. the same criterion *whenever* the function $f(x)$ satisfies the condition (as noted by Eccelston and Kiefer (1981))

$$f(cx) = a_c f(x)+b_c \text{ for some } a_c>0 \text{ corresponding to every } c>0.$$

In particular, if d is ψ_f–optimum in $\mathbf{D}^*(b,v,k,r)$, then $\tilde{d}$ is ψ_f–optimum in $\tilde{\mathbf{D}}^*(\tilde{b},\tilde{v},\tilde{k},\tilde{r})$. Thus the *Linked Block Designs* (**LBDs**) which are defined as duals of the **BBDs**, are found to be optimal within the class of equireplicate designs. The commonly used **A**–, **D**– and **E**–optimality criteria are all covered by the above form of f.

Sinha (1972), Shah *et al* (1976) and John and Mitchell (1977) had *all* independently studied some aspects of optimality of dual designs within the restricted class of equireplicate designs. The above presentation unifies the results so far known.

Remark 2.3.5. It is difficult to establish optimality of duals of optimal designs (viz., the **LBDs**) within the *unrestricted* class $\tilde{\mathbf{D}}(\tilde{b},\tilde{v},\tilde{k})$ of *all* connected designs. Cheng (1980) and Jacroux (1980, 1985) have made an attempt in this direction and have succeeded in establishing, among other things, **E**-optimality of the **LBDs** in the unrestricted class. These will be discussed in Chapter Three in the context of specific **A**–, **D**– and **E**-optimal designs.

REFERENCES

Cheng, C.S. (1978). Optimality of certain asymmetrical experimental designs. *Ann. Statist.*, *6*, 1239-1261.

Cheng, C.S. (1979). Optimal incomplete block designs with four varieties. *Sankhyā* (B), *41*, 1-14.

Cheng, C.S. (1980). On the *E*-optimality of some block designs. *J.R. Statist. Soc. B*, *42*, 205-209.

Conniffe, D. and Stone, J. (1974). The efficiency factor of a class of incomplete block designs. *Biometrika*, *61*, 633-636.

Conniffe, D. and Stone, J. (1975). Some incomplete block designs of maximum efficiency. *Biometrika*, *62*, 685-686.

Eccleston, J.A. and Kiefer, J. (1981). Relationships of optimality for individual factors of a design. *J. Statist. Planning and Inference, 5*, 213-219.

Fort, M.K. and Hedlund, G.A. (1958). Minimal covering by pairs of triplets. *Pacific J. Math., 8*, 709-719.

Gaffke, N. (1981). Some classes of optimality criteria and optimal designs for complete two-way layouts. *Ann. Statist., 9*, 893-898.

Jacroux, M. (1980). On the E-optimality of regular graph designs. *J.R. Statist. Soc. B, 42*, 205-209.

Jacroux, M. (1985). Some sufficient conditions for type-I optimality of block designs. *J. Statist. Planning and Inference, 11*, 385-394.

John, J.A. and Mitchell, T.J. (1977). Optimal incomplete block designs. *J.R. Statist. Soc. B, 39*, 39-43.

Kiefer, J. (1958). On the nonrandomized optimality and randomized nonoptimality of symmetrical designs. *Ann. Math. Statist., 29*, 675-699.

Roy, B.K. and Shah, K.R. (1984). On the optimality of a class of minimal covering designs. *J. Statist. Planning and Inference, 10*, 189-194.

Roy, J. (1958). On the efficiency factor of block designs. *Sankhyā*, 19, 181-188.

Shah, K.R., Raghavarao, D. and Khatri, C.G. (1976). Optimality of two and three factor designs. *Ann. Statist., 4*, 419-422.

Sinha, B.K. (1970). On the optimality of some designs. *Cal. Statist. Assoc. Bull., 19*, 1-22.

Sinha, B.K. (1972). *Contribution to comparison of experiments: Optimum experiments for linear inference*. Unpublished Ph.D. Thesis. Calcutta University.

Takeuchi, K. (1961). On the optimality of certain type of PBIB designs. *Rep. Statist. Res. Un. Japan Sci. Engrs., 8*, 140-145.

Takeuchi, K. (1963). A remark added to 'On the optimality of certain type of PBIB designs'. *Rep. Statist. Appl. Res. Un. Japan Sci. Engrs., 10*, 47.

Tocher, K.D. (1952). The design and analysis of block experiments (with discussion). *J.R. Statist. Soc. B, 14*, 45-100.

Yeh, C.M. (1986). Conditions for universal optimality of block designs. *Biometrika, 73*, 701-706.

Yeh, C.M. (1988). A class of universally optimal binary block designs. *J. Statist. Planning and Inference, 18*, 355-361.

CHAPTER THREE

BLOCK DESIGNS: SPECIFIC OPTIMALITY

1. Introduction

In Chapter Two, we primarily discussed a general optimality result for block designs. If in a given set-up $\mathbf{D}(b,v,k)$, a **BBD** or a **GDD** with $m=2$, $n=v/2$, $\lambda_2=\lambda_1+1$ exists, then it is optimal in a very general sense. However, in many situations, such designs do *not* exist, thereby rendering considerable difficulty in the search for optimal designs. This calls for further studies on the nature of optimal designs at least w.r.t. some specific optimality criteria. There is no denying the fact that a satisfactory study of optimality aspects of designs in a given set-up is quite difficult and, as yet, we do *not* have many results available. Our purpose in this Chapter is to present various results so far known w.r.t. the criteria of A–, D–, E– and MV–optimality. Indeed the A– and E–optimality criteria have received comparatively more attention of the researchers and the MV–optimality criterion the least.

We present the results in the following order. First, in section 2, we discuss the E–optimality criterion and related results. Next, in section 3, we present the notion of **Efficiency Factor** of a block design and follow it up by a detailed account of its applicability in deriving A–optimal designs. In section 4, we discuss results pertinent to MV–optimality. Next, we pass on to a study of D–optimal designs in section 5. In section 6 we deal with some conjectures related to the nature of optimal designs. In section 7 we give a very brief outline of the work done in the optimality studies for block designs with unequal block sizes. At the end, we have included a summary of the optimality results for more familiar block designs.

We exclude throughout the trivial design set-ups $\mathbf{D}(b,v,k)$ where $v\,|\,k$. Also we mostly work with set-ups where $v\,|\,bk$. Of course, we occasionally restrict to $k<v$. It must be noted that for $k>v$, the term *binary* is to be understood in the sense of *generalized binary* which means $n_{ij}=[k/v]$ or $[k/v]+1$ for all $1\le i\le v$, $1\le j\le b$.

Cheng (1981) pointed out an interesting relation between the C-matrix of a design with blocks of size k and that of a design in which every block of the original design is replaced by $\binom{k}{2}$ blocks consisting of the pairs formed out of the treatments in that block. Denoting the two C-matrices by $\mathbf{C}_k$ and $\mathbf{C}_2$ respectively, it follows that

$$2\mathbf{C}_2 = k\mathbf{C}_k \tag{3.1.1}$$

One may attempt to use this connection in the search for an optimal design. It is easy to see that if for a design in $\mathbf{D}(b,v,k)$, the corresponding design in $\mathbf{D}(b\binom{k}{2},v,2)$ is optimal, then the former design is also optimal in $\mathbf{D}(b,v,k)$. An advantage in this method is that in $\mathbf{D}(b\binom{k}{2},v,2)$, the search for an optimal design is necessarily confined to *strictly* binary designs. On the other hand, this method is *not* likely to be very useful. This is due to the following reasons.

(i) In most situations, a design which is optimal in $\mathbf{D}(b\binom{k}{2},v,2)$ will *not* correspond to a design in $\mathbf{D}(b,v,k)$. For example, the optimal design in $\mathbf{D}(210,21,2)$ is an unreduced **BIBD** while in the class $\mathbf{D}(14,21,6)$ there does *not* exist any **BIBD**, thereby violating (3.1.1). This is an extreme but certainly not an uncommon situation.

(ii) If a non-binary design happens to be optimal in $\mathbf{D}(b,v,k)$, the corresponding design in $\mathbf{D}(b\binom{k}{2},v,2)$ will *not* be optimal.

In general, there is no way of examining if an optimal design in $\mathbf{D}(b\binom{k}{2},v,2)$ corresponds to a *feasible* design in $\mathbf{D}(b,v,k)$. Even if it does, there does *not* seem to be an easy method to handle this problem of identification. Of course, as Cheng (1981) points out, in some cases, this consideration may simplify the proof of some known optimality results. We will *not* discuss this issue any further.

2. E-Optimal Designs

For given b,v and k, an **E**-optimal design seeks to maximize the minimum positive eigenvalue $x_{(1)}$ of the **C**-matrix within the class $\mathbf{D}(b,v,k)$. It is natural then to look for an *attainable* upper bound for $x_{(1)}$ or, equivalently, for $kx_{(1)}$. In the process, there have been suggested various bounds for various subclasses of designs and these have been compared to come up with an **E**-optimal design in the *entire* class $\mathbf{D}(b,v,k)$.

We treat the cases of $v \mid bk$ and $v \nmid bk$ separately in the subsections 2.1 and 2.2 respectively.

2.1. E-optimal designs for $v \mid bk$

It is generally believed that an **E**-optimal design will be binary and equireplicate. In view of this, the standard approach has been to first develop bounds for $kx_{(1)}$ (a) for *non-equireplicate* designs, (b) for *equireplicate non-binary* designs, and (c) for *equireplicate binary* designs. It may then be concluded that if there is an equireplicate binary design for which $kx_{(1)}$ exceeds each of the bounds under (a) and (b) above, the **E**-optimal design must necessarily be equireplicate and binary. Further to this, if for such a design of the latter type, $kx_{(1)}$ does, in fact, attain the bound under (c) above, then certainly it is **E**-optimal. This has been the basic strategy followed by Jacroux (1980a) and Cheng (1980) in deriving **E**-optimal designs. Most of the results given in this subsection are due to them.

We now present two Lemmas which are fundamental to the present study. These are essentially based on the type of reasoning first initiated by Takeuchi (1961). In the following modified form, these are to be found in Jacroux (1980a). Recall that every **C**-matrix has row and column sums zeroes.

Lemma 3.2.1. Define a symmetric matrix $\mathbf{T}_x = ((t_{xii'}))$ as

$$\mathbf{T}_x = k\mathbf{C}-x(\mathbf{I}_v - \mathbf{J}/v). \tag{3.2.1}$$

If for some $x>0$, $\mathbf{T}_x$ has (i) at least one negative eigenvalue, or (ii) at least two eigenvalues as zero, then $kx_{(1)} \leq x$.

Proof. It is enough to observe that the eigenvalues of $\mathbf{T}_x$ are 0, $kx_{(1)}-x \leq kx_{(2)}-x \leq \cdots \leq kx_{(v-1)} - x$.

Remark 3.2.1. If for some $x>0$,

$$t_{xii} = t_{xi'i'} = \pm t_{xii'} \text{ for some } i \neq i' \tag{3.2.2}$$

then condition (ii) above holds and, hence, $kx_{(1)} \leq x$.

Lemma 3.2.2. If for some i, $t_{x_0ii} \leq 0$ for a suitable choice of x, say $x = x_0$, then $kx_{(1)} \leq x_0$.

Proof. If $t_{x_0ii}<0$, then clearly $kx_{(1)}<x_0$. If $t_{x_0ii} = 0$, then either $t_{x_0is} = 0$ for all s or $t_{x_0is} \neq 0$ for some s. In the first case, 0 is an eigenvalue of multiplicity at least two (since $\mathbf{T}_{x_0}$ has row and column sums zeroes). In the second case, $\mathbf{T}_{x_0}$ has a negative eigenvalue. In either case, $kx_{(1)} \leq x_0$.

Below we first develop bounds for non-equireplicate designs. For such a design $r_{(1)}<r$ where $r_{(1)} \leq r_{(2)} \leq \cdots \leq r_{(v)}$ and $r = bk/v$. We use the following notations.

$$\left.\begin{array}{rl}
\text{(i)} & \alpha = [k/v] = [r/b] \\
\text{(ii)} & l = r-b\alpha, \ r = l(\alpha+1) + (b-l)\alpha \\
\text{(iii)} & \theta = l(\alpha+1)^2 + (b-l)\alpha^2 \\
\text{(iv)} & rk-\theta = \delta(v-1) + \epsilon, \ 0 \leq \epsilon < v-1 \\
\text{(v)} & \alpha_1 = [r_{(1)}/b] \\
\text{(vi)} & l_1 = r_{(1)} - b\alpha_1, \ r_{(1)} = l_1(\alpha_1+1) + (b-l_1)\alpha_1 \\
\text{(vii)} & \theta_1 = l_1(\alpha_1+1)^2 + (b-l_1)\alpha_1^2. \\
\text{(viii)} & \lambda_{11} = \sum_j n_{1j}^2 \text{ with } \sum_j n_{1j} = r_{(1)}
\end{array}\right\} \tag{3.2.3}$$

We will also be frequently using the functions g_u and h_u defined below.

$$g_u = rk-\theta+u, \ h_u = v(rk-\theta-u)/(v-2). \tag{3.2.4}$$

Now observe that $\lambda_{11} \geq \theta_1$. Next it may be verified that

$$\theta - \theta_1 \leq (2\alpha+1)(r - r_{(1)}). \tag{3.2.5}$$

Further, $\alpha = [k/v]$ implies $k \geq v\alpha+1$ (as $v \nmid k$) and, hence, $k \geq 2\alpha+3$ for $k \geq 3$, $v \geq 4$. This together with (3.2.5) implies that

$$\theta - \theta_1 \leq (k-2)(r - r_{(1)}) \text{ for } k \geq 3,\ v \geq 4. \tag{3.2.6}$$

We are now in a position to enunciate the following.

Proposition 3.2.1. For any non-equireplicate design in $\mathbf{D}(b,v,k)$,

$$\text{(1)}\ kx_{(1)} < g_{\delta-1} = rk - \theta + \delta - 1 \text{ for } k \geq 3,\ v \geq 4 \tag{3.2.7}$$

$$\text{(2)}\ kx_{(1)} \leq \frac{(r-1)(k-1)v}{v-1} \tag{3.2.8}$$

Proof of (1). Set $x_0 = v(r_{(1)}k - \lambda_{11})/(v-1)$ in (3.2.1) and verify that $t_{x_0 11} = 0$. Applying Lemma 3.2.2, we then have

$$\begin{aligned} kx_{(1)} &\leq x_0 = v(r_{(1)}k - \lambda_{11})/(v-1) \leq v(r_{(1)}k - \theta_1)/(v-1) \text{ as } \lambda_{11} \geq \theta_1 \\ &\leq v\{r_{(1)}k + (r - r_{(1)})(k-2) - \theta\}/(v-1) \text{ by (3.2.6)} \\ &= v\{(rk-\theta) - 2(r - r_{(1)})\}/(v-1). \end{aligned}$$

As $r_{(1)} < r$, this gives

$$\begin{aligned} kx_{(1)} &\leq v(rk - \theta - 2)/(v-1) \\ &= v\{\delta(v-1) + \epsilon - 2\}/(v-1) \text{ by (3.2.3) (iv)} \\ &= \delta v + \frac{v(\epsilon-2)}{v-1} \\ &< \delta v + (\epsilon - 1) = (\delta - 1) + (v-1)\delta + \epsilon \\ &= (rk - \theta) + (\delta - 1) = g_{\delta-1}. \end{aligned} \tag{3.2.9}$$

This proves (1).

Proof of (2). Set $x_0 = \dfrac{(r-1)(k-1)v}{v-1}$ in (3.2.1) and note that

$$t_{x_0 11} = kr_{(1)} - \lambda_{11} - (r-1)(k-1)$$

$$\leq kr_{(1)} - \theta_1 - (r-1)(k-1).$$

Now using (3.2.3) (vi), (vii), it can be verified that $kr_{(1)} - \theta_1 \leq (r-1)(k-1)$ so that $t_{x_0 11} \leq 0$. Hence the result follows by an application of Lemma 3.2.2.

It may be noted that this inequality does *not* require the condition $k \geq 3$ and/or $v \geq 4$. Further, it holds whether or not the design is binary.

Remark 3.2.2. Even though two bounds have been presented above, as a matter of fact, it is enough that for an equireplicate binary design, $kx_{(1)}$ exceeds the smaller of the two. In this context, Cheng (1980) has observed that for $3 \leq k < v$, the bound in (3.2.8) does *not* exceed that in (3.2.7). For $k = 2$, (3.2.8) serves as a useful bound to $kx_{(1)}$. This case will be discussed towards the end.

Our next inequality relates to equireplicate non-binary designs.

Proposition 3.2.2. For any equireplicate non-binary design, $kx_{(1)} < g_{\delta-1}$.

Proof. As the design is *not* binary, $n_{ij} < \alpha$ or $> \alpha+1$ for *some* (i,j), $1 \leq i \leq v$, $1 \leq j \leq b$. This gives $\lambda_{ii} \geq \theta+2$ for *some* i. Set, now, $x_0 = v(rk-\theta-2)/(v-1)$ in (3.2.1). Then $t_{x_0 ii} = (rk-\lambda_{ii}) - (rk-\theta-2) \leq 0$. Hence, by an application of Lemma 3.2.2, we have

$$kx_{(1)} \leq x_0 = v(rk-\theta-2)/(v-1) \text{ which is (3.2.9).}$$

The rest is clear. Hence, the proposition follows.

Remark 3.2.3. It is clear that $g_{\delta-1}$ serves as an upper bound for $kx_{(1)}$ for *any* design which fails to be binary and equireplicate, for $k \geq 3$.

The next inequality is quite general in nature and it deals with designs which are binary *and* equireplicate.

Proposition 3.2.3. For an equireplicate binary design,

$$(1)\quad kx_{(1)} \leq rk-\theta+\lambda_{ii'} \text{ for } all \;\; i \neq i' \tag{3.2.10}$$

with strict inequality when $\lambda_{is} \neq \lambda_{i's}$ for *some* $s \neq i, \neq i'$;

$$(2)\quad kx_{(1)} \leq v(rk-\theta-\lambda_{ii'})/(v-2) \text{ for all } i \neq i' \tag{3.2.11}$$

with strict inequality when $(\lambda_{is}+\lambda_{i's})/2 \neq (rk-\theta-\lambda_{ii'})/(v-2)$ for *some* $s \neq i, \neq i'$.

Proof. First observe that λ_{ii}'s are *all* equal as the underlying designs are binary and equireplicate. Set, now, $x = rk-\theta+\lambda_{ii'}$ (respectively, $x = v(rk-\theta-\lambda_{ii'})/(v-2)$) and observe that $t_{xii} = t_{xi'i'} = t_{xii'}$ (respectively, $t_{xii} = t_{xi'i'} = -t_{xii'}$). Then referring to Lemma 3.2.1 and Remark 3.2.1 following it, the two results follow.

That the inequalities are strict follows from an examination of singularity or otherwise of the 3×3 symmetric submatrix

$$\begin{bmatrix} t_{xii} & t_{xii'} & t_{xis} \\ & t_{xi'i'} & t_{xi's} \\ & & t_{xss} \end{bmatrix}.$$

Remark 3.2.4. For a non-binary equireplicate design, we already have seen that (3.2.9) holds i.e., $kx_{(1)} \le (rk-\theta-2)v/(v-1)$. Since $\theta \ge r$, it follows that $kx_{(1)} \le \{r(k-1)-2\}v/(v-1)$. Further, as noted earlier, for a non-binary non-equireplicate design, $kx_{(1)} \le (r-1)(k-1)v/(v-1)$. Thus, for $k \ge 3$, $kx_{(1)} \le \{r(k-1)-2\}v/(v-1)$ for *any* non-binary design as this bound exceeds the other in this case.

Corollary 3.2.1. For a binary equireplicate design in $\mathbf{D}(b,v,k)$,

$$kx_{(1)} \le g_u, \quad kx_{(1)} \le h_\omega \tag{3.2.12}$$

where

$$u = \min_{i<i'} \lambda_{ii'} \text{ and } \omega = \max_{i<i'} \lambda_{ii'} \tag{3.2.13}$$

Since $g_x(h_x)$ is increasing (decreasing) in x, it is clear that

$$\left.\begin{aligned} kx_{(1)} &\le g_\delta = h_\delta \text{ for } \epsilon = 0 \\ kx_{(1)} &\le \min(g_\delta, h_{\delta+1}) \text{ for } \epsilon > 0 \end{aligned}\right\} \tag{3.2.14}$$

for *any* binary equireplicate design.

This is because for $\epsilon = 0$, $u \le \delta \le \omega$ while for $\epsilon > 0$, $u \le \delta < \delta+1 \le \omega$. It can easily be seen that $g_\delta \underset{<}{>} h_{\delta+1}$ as $\epsilon \underset{>}{<} v/2$. Accordingly, two cases emerge e.g., $kx_{(1)} \le g_\delta \le h_{\delta+1}$ for $\epsilon \ge v/2$ and $kx_{(1)} \le h_{\delta+1} < g_\delta$ for $\epsilon < v/2$. Moreover, for $\epsilon = 0$, it can be checked that $g_\delta = h_\delta$ and $kx_{(1)}$ reaches this bound iff the design is a **BBD**.

The above propositions can now be applied to establish **E**-optimality of some classes of designs. The first result in this direction (due originally to Takeuchi (1961, 1963) is given in the following.

Theorem 3.2.1. For $k \ge 3$ and $v \ge 4$, a design for which $kx_{(1)} = g_\delta$ is necessarily binary and equireplicate, and it is **E**-optimal in $\mathbf{D}(b,v,k)$. Further, such a design is necessarily either a **BBD** or a **GDD** with $\lambda_2 = \lambda_1 + 1$.

Proof. Since $g_{\delta-1}$ is an upper bound to $kx_{(1)}$ for *any* design which is either non-binary or non-equireplicate, and since $g_{\delta-1} = g_\delta - 1$, the first part of the Theorem is evident.

The second part provides a characterization of such **E**-optimal designs. This we demonstrate as follows.

It is easy to verify that when a **BBD** exists in $\mathbf{D}(b,v,k)$, then the non-zero eigenvalues of its **C**-matrix are *all* equal to g_δ. If, on the other hand, a **BBD** does *not* exist but a **GDD** with $\lambda_2 = \lambda_1 + 1$ exists, then it is easily seen that for such a design, $kx_{(1)} = g_\delta$. For the converse part i.e., to see that these are the only possibilities, suppose a design has $kx_{(1)} = g_\delta$ and it is *not* a **BBD**. We will show that it must be a **GDD** with $\lambda_2 = \lambda_1 + 1$.

Not to obscure the essential steps of reasoning, we argue as follows:

(i) Since $kx_{(1)} = g_\delta$, the design is necessarily binary *and* equireplicate. Consequently, the diagonal elements of the **C**-matrix are all equal to $(rk-\theta)/k$. Further, (3.2.12) implies that $u = \delta$ or, otherwise, $kx_{(1)}<g_\delta$ which is *not* the case. Suppose that $\lambda_{12} = \delta$. It is seen from the proof of Proposition 3.2.3 that for $x = rk-\theta+\delta$, this yields

$$t_{x11} = t_{x22} = t_{x12} = (rk-\theta) - (rk-\theta+\delta)(\frac{v-1}{v})$$

$$= \{rk-\theta-\delta(v-1)\}/v = \epsilon/v \text{ (by (3.2.3) (iv))} = t \text{ (say)}, \; 0<t<1.$$

(ii) Next it is evident that $t_{xii} = t$ for *all* i. If now $\lambda_{ii'} = \delta+l$ for some $l \geq 1$, then $t_{xii'} = t-l$. Since $kx_{(1)} = g_\delta$, $\mathbf{T}_x$ in (3.2.1) with $x = rk-\theta+\delta = g_\delta$ must be nnd. This means that l *cannot* exceed 1 as, otherwise, a (2×2) principal minor $\begin{bmatrix} t & t-l \\ t-l & t \end{bmatrix}$ would produce a negative determinant. Hence, $l = 0$ or 1 are the only possibilities. This leads to $\lambda_{ii'} = \delta$ or $\delta+1$ and since $rk-\theta = \sum_{i'(\neq i)} \lambda_{ii'} = \delta(v-1) + \epsilon$, precisely ϵ of the $\lambda_{ii'}$'s are *each* equal to $\delta+1$ for *every* i, $1\leq i\leq v$. Thus every row of $\mathbf{T}_x$ has the element t occurring $(v-\epsilon)$ times and the element $t-1$ occurring ϵ times.

(iii) Set $t_{x11} = t_{x12} = \cdots = t_{x1(v-\epsilon)} = t$, $t_{x1(v-\epsilon+1)} = \cdots = t_{x1v} = t-1$. Clearly, $t_{x21} = t_{x22} = t$. Further, since in Proposition 3.2.3 (part (1)), *strict inequality* does *not* hold with $i = 1$, $i' = 2$ (as we assumed $\lambda_{12} = \delta$), we must have $t_{x1s} = t_{x2s}$ for all s. The same argument shows that the first $(v-\epsilon)$ rows of $\mathbf{T}_x$ are identical and produce the submatrix $[t\mathbf{J}_{(v-\epsilon)\times(v-\epsilon)}|(t-1)\mathbf{J}_{(v-\epsilon)\times\epsilon}]$.

(iv) Continuing this argument in an analogous fashion, we end up with the following *formal* structure of $\mathbf{T}_x$:

$$\mathbf{T}_x = \begin{bmatrix} t\mathbf{J} & (t-1)\mathbf{J} & \cdots & (t-1)\mathbf{J} \\ & t\mathbf{J} & \cdots & (t-1)\mathbf{J} \\ & & & \\ & & \cdots & t\mathbf{J} \end{bmatrix}$$

This dictates a similar structure for the **C**-matrix which shows that the underlying design can be identified as a **GDD** with $n = v-\epsilon$, $m = \dfrac{v}{v-\epsilon} = \dfrac{v}{n}$ and

$\lambda_2 = \lambda_1 + 1$.

This completes the proof.

Remark 3.2.5. Clearly, for $\epsilon>0$, $m\geq 2$ and this yields $\epsilon \geq v/2$ which in its turn implies $g_\delta \leq h_{\delta+1}$. Thus one of the two cases is taken care of. The next theorem deals with the case of $0<\epsilon<v/2$.

Theorem 3.2.2. For given (b,v,k) with $k\geq 3$ and $v\geq 4$, suppose $0<\epsilon<v/2$. Then a design for which $kx_{(1)} = h_{\delta+1}$ is necessarily binary and equireplicate, and it is **E**-optimal in $\mathbf{D}(b,v,k)$. Further, for such a design the resulting $\mathbf{T}_x$-matrix with $x = h_{\delta+1}$ necessarily assumes the form of a block diagonal matrix with components given by $\begin{pmatrix} \mathbf{J} & -\mathbf{J} \\ -\mathbf{J} & \mathbf{J} \end{pmatrix}$ of orders $2p_1, 2p_2, \ldots$ where p_i's satisfy $p_1+p_2+ \cdots = v/2$. In particular, the **GDDs** with $n = 2$, $\lambda_1 = \lambda_2+1$ (>1) as also the **GDDs** with $m = 2$, $\lambda_2 = \lambda_1+2$ are **E**–optimal.

Proof. First observe that $\epsilon>0$ implies $g_{\delta-1}<h_{\delta+1}$. Hence the first part of the Theorem follows. The second part on characterization of the form of $\mathbf{T}_x$ needs close arguments which are developed below.

Since $kx_{(1)}$ attains the bound $h_{\delta+1}$, it is clear that $\omega = \delta + 1$. Suppose $\lambda_{12} = \delta + 1$. Then, for $x = h_{\delta+1} = v(rk-\theta-\delta-1)/(v-2)$,

$$t_{x11} = t_{x22} = -t_{x12} = \frac{v-1-\epsilon}{v-2} = t' \text{ (say)}, \ 0<t'\leq 1.$$

Referring to (3.2.11), since $\mathbf{T}_x$ is nnd, we must have, for every $s \neq 1, \neq 2$, $\lambda_{1s}+\lambda_{2s} = 2(rk-\theta-\lambda_{12})/(v-2) = 2(rk-\theta-\delta-1)/(v-2) = 2\delta+2(\epsilon-1)/(v-2)$. As $\epsilon<v/2$, $\lambda_{1s}+\lambda_{2s}<2\delta+1$ so that $\lambda_{1s} = \lambda_{2s} = \delta$ for every $s \neq 1, \neq 2$. This gives $\epsilon = 1$ and, hence, $t' = 1$. Thus, all diagonal elements of $\mathbf{T}_x$ are unity. As $\mathbf{T}_x$ is nnd and all its off-diagonal elements are integers, these are necessarily confined to $(0,\pm 1)$ which, in their turn, imply that $\lambda_{ii'}$'s assume values $(\delta,\delta\pm 1)$. As regards the structure of $\mathbf{T}_x$, we have the following information:

(i) $\mathbf{T}_x$ is nnd with $t_{xii} = 1$ for all i.

(ii) $t_{xii'} = 0, \pm 1$ for all $i \neq i'$.

(iii) $\sum_{i'=1}^{v} t_{xii'} = 0$ for all i, $t_{x12} = -1$ (assumed)

(iv) $t_{xii'}=-1 \Rightarrow t_{xis}+t_{xi's} = 0$, i.e., $t_{xis} = t_{xi's} = 0$ or $t_{xis} = \pm 1$, $t_{xis'} = \mp 1$ for all $s \neq i, \neq i'$.

Without any loss, set now $t_{x1s} = 1$, $t_{x1s'} = -1$ for some (s,s'), $s \neq s' \neq 2$. Then we immediately deduce the following structure for the 4×4 submatrix of $\mathbf{T}_x$ formed of rows and columns numbered $(1,2,s,s')$:

$$\begin{bmatrix} 1 & -1 & 1 & -1 \\ -1 & 1 & -1 & 1 \\ 1 & -1 & 1 & -1 \\ -1 & 1 & -1 & 1 \end{bmatrix} \text{ which is equivalent (up to permutation) to } \begin{bmatrix} \mathbf{J}_2 & -\mathbf{J}_2 \\ -\mathbf{J}_2 & \mathbf{J}_2 \end{bmatrix}.$$

Moreover, if now $t_{x1s''} = 0$ for some $s'' \neq s, \neq s', \neq 2$, then we further deduce that $t_{x2s''} = t_{xss''} = t_{xs's''} = 0$. Thus, starting with the first entry t_{x11} of $\mathbf{T}_x$, we end up with a block diagonal matrix of the form $\begin{pmatrix} \mathbf{J} & -\mathbf{J} \\ -\mathbf{J} & \mathbf{J} \end{pmatrix}$. Certainly, this argument can be carried further starting with some h *not* covered by the above block diagonal and observing that $t_{xhh} = 1$ while $t_{xih} = 0$ for every i covered by the above. This leads to block diagonals with components given by $\begin{pmatrix} \mathbf{J} & -\mathbf{J} \\ -\mathbf{J} & \mathbf{J} \end{pmatrix}$.

The last part of the Theorem follows from the observation that two extreme forms of $\mathbf{T}_x$ are given by $\begin{pmatrix} 1 & -1 \\ -1 & 1 \end{pmatrix} \otimes \mathbf{I}_{v/2}$ and $\begin{pmatrix} \mathbf{J}_{v/2} & -\mathbf{J}_{v/2} \\ -\mathbf{J}_{v/2} & \mathbf{J}_{v/2} \end{pmatrix}$ where $\otimes$ denotes the operator of Kronecker product. These correspond respectively to the **GDDs** with $n = 2$, $\lambda_1 = \lambda_2+1$ (>1, as otherwise the **GDD** is disconnected) and the **GDDs** with $m = 2$, $\lambda_2 = \lambda_1+2$. Jacroux (1983a) used different arguments to establish **E**-optimality of the **GDDs** with $m = 2$, $\lambda_2 = \lambda_1+2$. Such **GDDs** seem to be rather rare for $k>2$.

It must be noted that, in general, the above characterization results in a structure of $\mathbf{T}_x$ with off-diagonal elements $0, \pm1$ so that the resulting **E**-optimal design will involve at most three distinct concurrences $(\delta, \delta\pm1)$. Accordingly, if $\lambda_{ii'}$'s assume three distinct values $(\delta, \delta\pm1)$, the underlying designs may be termed *most-balanced 3-concurrence*. On the other hand, the above **GDDs** refer to situations where the $\mathbf{T}_x$-matrix has either $(0,-1)$ or (±1) as its off-diagonal elements. A most balanced 3-concurrence **E**-optimal design possesses the following group structure of the v treatments.

The treatments fall into t (>1) groups with $2p_s$ treatments in the sth group with $p_s>1$ for at least one s and $\sum_1^t p_s = v/2$. Divide the treatments of the sth group into two sets $\mathbf{G}_s$ and $\overline{\mathbf{G}}_s$ each having p_s treatments. Then $\mathbf{G} = \cup(\mathbf{G}_s \cup \overline{\mathbf{G}}_s)$ is the set of all v treatments. As regards $\lambda_{ii'}$'s, we have that

$$\begin{aligned} \lambda_{ii'} &= \delta-1 \text{ for both } i, i' \epsilon \mathbf{G}_s \text{ or } \overline{\mathbf{G}}_s,\ i \neq i' \\ &= \delta+1 \text{ for } i \epsilon \mathbf{G}_s,\ i' \epsilon \overline{\mathbf{G}}_s \text{ or the reverse} \\ &= \delta \text{ for } i \epsilon \mathbf{G}_s \cup \overline{\mathbf{G}}_s;\ i' \epsilon \mathbf{G}_{s'} \cup \overline{\mathbf{G}}_{s'},\ s \neq s'. \end{aligned}$$

Such designs form very special subclasses of what are generally termed *Intra- and Inter-Group Balanced Block Designs* (**IIGBBDs**). (Vide Rao (1947)). In the literature, combinatorial and constructional aspects of such designs with unequal replications have been studied quite extensively. See, for example, Adhikary (1965). Below we give an example of an **E**-optimal 3-concurrence **IIGBBD** with $\lambda_{ii'} = 0,\ 1$ or 2.

Example. $b = v = 12,\ r = k = 4,\ \mathbf{G}_1 = (1,2),\ \overline{\mathbf{G}}_1 = (3,4),\ \mathbf{G}_2 = (5),\ \overline{\mathbf{G}}_2 = (6),\ \mathbf{G}_3 = (7),$ $\overline{\mathbf{G}}_3 = (8),\ \mathbf{G}_4 = (9),\ \overline{\mathbf{G}}_4 = (10),\ \mathbf{G}_5 = (11),\ \overline{\mathbf{G}}_5 = (12).$

Blocks	Treatments				Blocks	Treatments				Blocks	Treatments			
1	1	2	5	6	5	2	4	9	12	9	5	7	9	11
2	1	2	7	8	6	2	4	10	11	10	5	7	10	12
3	1	3	9	10	7	3	4	5	8	11	6	8	9	10
4	1	3	11	12	8	3	4	6	7	12	6	8	11	12

It may be noted that a **GDD** with $m = 6,\ n = 2,\ \lambda_1 = 2,\ \lambda_2 = 1$ also exists in this design set-up. The above study is due to Sinha and Shah (1988).

Remark 3.2.6. The following chart exhibits the scope of applications of the above Theorems in terms of values assumed by ϵ for given b,v,k, and the nature of **E**-optimal designs as derived therefrom.

	Values of ϵ: 0	1	$v/2$	s.t. $v/(v-\epsilon)$ is an integer
Value of $kx_{(1)}$ attained	$h_\delta = g_\delta$	$h_{\delta+1}(<g_\delta)$	$h_{\delta+1}(=g_\delta)$	$g_\delta(<h_{\delta+1})$
Nature of **E**-optimal designs	**BBD**	**GDD** with $n = 2,\ \lambda_1 = \lambda_2+1$ or **GDD** with $m = 2,\ \lambda_2 = \lambda_1+2$ or an *appropriate* **IIGBBD**.	**GDD** with $m = 2,\ \lambda_2 = \lambda_1+1$	**GDD** with $m = \dfrac{v}{v-\epsilon},\ \lambda_2 = \lambda_1+1$

The optimality results so far discussed in this and the previous chapter mainly relate to the **BBDs** and to some classes of the **GDDs**. The essential features of these designs are the following:

(i) $n_{ij} = [k/v]$ or $[k/v] + 1$ for all $1 \leq i \leq v,\ 1 \leq j \leq b$;

(ii) r_i's are all equal;

(iii) $|\lambda_{ii'} - \lambda_{uu'}| \leq 1$ for all $i \neq i',\ u \neq u'$.

As noted earlier in Chapter Two (section 2), (i) and (ii) imply that the λ_{ii}'s are all equal.

Designs satisfying (i), (ii) and (iii) above have been called *Regular Graph Designs* (**RGDs**) by John and Mitchell (1977). Various optimality aspects of the **RGDs** have been studied in the literature. We present additional results on E-optimality in this section. Results relating to other optimality criteria are presented in sections 3-5. We hope that this study will help in appreciating the nature of a couple of conjectures regarding the affirmative role of the **RGDs** in our search for optimal designs. These conjectures are briefly discussed in section 6.

Referring back to the chart of E-optimal designs presented above, we note that for any of the values of ϵ covered, the corresponding design (displayed below it) may *not* necessarily exist. Moreover, the chart does *not* cover many values of ϵ, in the range $0<\epsilon\leq v-1$. To tackle all such situations, the following observation due to Jacroux (1980a) may be helpful in the sense of restricting the search to an appropriate subclass of $\mathbf{D}(b,v,k)$.

Theorem 3.2.3. If for given b,v,k with $k\geq 3$, $v\geq 4$, there exists an **RGD** with $kx_{(1)}\geq \max(g_{\delta-1}, h_{\delta+2})$, then there exists an E-optimal **RGD**. Further, if strict inequality holds above, an E-optimal design is necessarily an **RGD**.

Proof. First note that a design may fail to be an **RGD** for any of the following reasons: (i) non-binary, (ii) binary non-equireplicate, (iii) binary equireplicate with $\lambda_{ii'}<\delta$ for some $i\neq i'$ and (iv) binary equireplicate with $\lambda_{ii'}>\delta+1$ for some $i\neq i'$. In cases (i) - (iii), $kx_{(1)}\leq g_{\delta-1}$ while in case (iv), $kx_{(1)}\leq h_{\delta+2}$. These are consequences of Propositions 3.2.1 - 3.2.3. Hence the Theorem.

Cheng (1980) and Jacroux (1980a) have listed some **RGDs** (which are, in fact, two associate **PBIBDs**) satisfying the conditions of the above Theorem.

We now discuss the case of $k=2$ and present the optimality results as far as available. We omit the proofs.

(a) The statement of Theorem 3.2.1 remains valid and a **BBD** or a **GDD** with $\lambda_2=\lambda_1+1$ is E-optimal. To see this, we first observe that a design for which $kx_{(1)}=g_\delta$ must necessarily be equireplicate as $g_\delta>\dfrac{v(r-1)}{v-1}$ which is the upper bound in (3.2.8) for $kx_{(1)}$ for *any* non-equireplicate design. The rest of the argument is clear.

(b) The statement of Theorem 3.2.2 also remains essentially valid in the sense that a **GDD** with $n=2$, $\lambda_1=\lambda_2+1$ has $kx_{(1)}=h_{\delta+1}$ and it is E-optimal. However, there remains the possibility of a non-equireplicate design also attaining this bound as $h_{\delta+1}=v(r-1)/(v-1)$. For such a design, of course, the replication numbers assume the values r, $r\pm 1$.

(c) A version of Theorem 3.2.3 in the case of $k=2$ spells out that existence of an **RGD** with

$$kx_{(1)}\geq \max(g_{\delta-1}, h_{\delta+2}, \frac{v(r-1)}{v-1})$$

implies existence of an **E**-optimal **RGD**. Cheng (1980) applied this technique for $k=2$ and observed that two-associate class **PBIBDs** with cyclic scheme, and with $\lambda_1=\lambda_2+1$ or $\lambda_1=\lambda_2-1$ and $v=5$, are **E**-optimal.

For *small* designs i.e., for designs with $rk<v$, the following Theorem due to Constantine (1982) is found to be useful in establishing optimality of some **RGDs** and, specifically, of some 2-associate class **PBIBDs**.

Theorem 3.2.4. Suppose for a design in $\mathbf{D}(b,v,k)$ with $k<v$, $kx_{(1)}=\dfrac{v(r-1)(k-1)}{v-k}$. Then it is **E**-optimal in $\mathbf{D}(b,v,k)$. Further, $rk<v$.

Proof. We give the proof in three steps.

Step I. Suppose for a given *equireplicate* design, a certain block, say, the first block consists of m distinct treatments, say, $(1,2,\ldots,m)$. Clearly, $2\le m\le k$. Consider the $m\times m$ principal minor of $k\mathbf{C}$ formed by these m rows and columns.

Let $\mathbf{x}=\begin{bmatrix}(1-\frac{m}{v})\mathbf{1}_m\\ (-\frac{m}{v})\mathbf{1}_{v-m}\end{bmatrix}$ so that $\mathbf{x'1}=0$ and $\mathbf{x'x}=\dfrac{m(v-m)}{v}$. Then, $kx_{(1)}$ being the smallest positive eigenvalue of $k\mathbf{C}$, we must have

$$(kx_{(1)})(\mathbf{x'x})\le\mathbf{x}'(k\mathbf{C})\mathbf{x}=\begin{pmatrix}\mathbf{1}_m\\ \mathbf{0}_{v-m}\end{pmatrix}'(k\mathbf{C})\begin{pmatrix}\mathbf{1}_m\\ \mathbf{0}_{v-m}\end{pmatrix}$$

$$=mrk-\sum_{i=1}^{m}\sum_{j=1}^{b}n_{ij}^2-\sum_{i\neq i'}^{m}\sum\lambda_{ii'}.$$

Now $\sum_{j=1}^{b}n_{ij}^2=\sum_{j=2}^{b}n_{ij}^2+n_{i1}^2\ge\sum_{j=2}^{b}n_{ij}+n_{i1}^2=n_{i1}^2+r-n_{i1}$. Therefore, $\sum_{i=1}^{m}\sum_{j=1}^{b}n_{ij}^2\ge\sum_{i=1}^{m}n_{i1}^2+mr-\sum_{j=1}^{m}n_{i1}=\sum_{i=1}^{m}n_{i1}^2+mr-k$. Further, $\lambda_{ii'}\ge n_{i1}n_{i'1}$. This gives

$$(kx_{(1)})\frac{m(v-m)}{v}\le mrk-\sum_{i=1}^{m}n_{i1}^2-mr+k-\sum_{i\neq i'}^{m}\sum n_{i1}n_{i'1}$$

$$=mrk-mr+k-(\sum_{i=1}^{m}n_{i1})^2=mrk-mr+k-k^2$$

$$=(mr-k)(k-1)$$

This yields

$$kx_{(1)}\le\frac{v(mr-k)(k-1)}{m(v-m)}=g^*(m)\text{ (say)}\tag{3.2.15}$$

Step II It is easy to verify that $g^*(m)$ is increasing in m so that

$$g^*(m) \le g^*(k) = \frac{v(r-1)(k-1)}{v-k}$$

which is the value stated in the Theorem.

Step III It is now enough to show that $g^*(k)$ exceeds the value of $kx_{(1)}$, corresponding to *any* non-equireplicate design. Using the bound in (3.2.8) (Vide Proposition 3.2.1), we know that for *any* non-equireplicate design, $kx_{(1)} \le \frac{v(r-1)(k-1)}{v-1}$ which is *strictly* less than $g^*(k)$. Hence the Theorem.

Remark 3.2.7. A careful examination of the case of equality in Step I with $m = k$ reveals that such an **E**-optimal design must be binary with $\lambda_{ii'} = 0$ or 1 for all $i \ne i'$. Further, Step III implies that such an **E**-optimal design must necessarily be equireplicate. This gives $r(k-1) = \sum_{i'(\ne i)} \lambda_{ii'} < v-1$ so that in the representation $r(k-1) = \delta(v-1) + \epsilon$, we have $\delta = 0$.

Remark 3.2.8. Note that earlier we had developed two bounds to $kx_{(1)}$ for binary equireplicate designs. These are given by (Vide (3.2.4) and (3.2.14)) $g_\delta = r(k-1) + \delta$, $h_{\delta+1} = \frac{v}{v-2}\{r(k-1)-\delta-1\}$. Clearly, since $g^*(k)$ is reached, we must have $g^*(k) \le \min(g_\delta, h_{\delta+1})$. This, in turn, reduces to the conditions $\delta = 0$ and $\epsilon < v - \frac{v}{k}$ i.e., $r(k-1)<v(k-1)/k$ i.e., to $rk<v$.

Thus the above Theorem seems to be applicable for *small* designs in the sense that $rk<v$. Constantine (1982) has listed some 2-associate class **PBIBDs** satisfying the conditions of the above Theorem. We mention in passing that the verification of $kx_{(1)} = v(r-1)(k-1)/(v-k)$ needs computation of the minimum positive eigenvalue of the **C**-matrix as *no simple structure* is as such inherent to such **E**-optimal designs.

We conclude this subsection by discussing the **E**-optimality aspects of duals of **E**-optimal designs. We will use the same notations $\tilde{\mathbf{D}}(\tilde{b},\tilde{v},\tilde{k})$ as used before in subsection 3.5 of Chapter Two. Recall that we have already established optimality (w. r. t. various optimality criteria, including **E**-optimality) of duals of *equireplicate* optimal designs in the *restricted* subclass of $\tilde{\mathbf{D}}(\tilde{b},\tilde{v},\tilde{k})$ comprising of equireplicate designs only. With the additional tools now available, it is possible to strengthen those results in the sense of extending them to the *entire* class $\tilde{\mathbf{D}}(\tilde{b},\tilde{v},\tilde{k})$. Clearly, one only needs to *dominate* over non-equireplicate designs. This may be accomplished by referring to Proposition 3.2.1. For $\tilde{k}\ge 3$, since the analogue of (3.2.8) viz., $(\tilde{r}-1)(\tilde{k}-1)\tilde{v}/(\tilde{v}-1)$ does *not* exceed the analogue of (3.2.7) in $\tilde{\mathbf{D}}(\tilde{b},\tilde{v},\tilde{k})$, it is enough to verify that the dual of an equireplicate binary **E**-optimal design provides, for its **C**-matrix,

$$\tilde{k}\tilde{x}_{(1)} \ge \frac{(\tilde{r}-1)(\tilde{k}-1)\tilde{v}}{(\tilde{v}-1)} = \frac{(r-1)(k-1)b}{(b-1)}.$$

However, it is interesting to note that $\tilde{k}\tilde{x}_{(1)} = kx_{(1)}$. So it amounts to verifying that $kx_{(1)} \ge (r-1)(k-1)b/(b-1)$. Once this is achieved, the *earlier* result on **E**-optimality of dual designs gets immediately extended to the entire class $\tilde{\mathbf{D}}(\tilde{b},\tilde{v},\tilde{k})$.

Cheng (1980), Jacroux (1980a) and Constantine (1982) have precisely put forward these arguments and succeeded in establishing over-all **E**-optimality of duals of (i) the **BBDs**, (ii) the **GDDs** with $\lambda_2 = \lambda_1 + 1$ and (iii) the **GDDs** with $\lambda_1 = \lambda_2 + 1$ (>1) and group size 2. Some other results on optimality of dual designs have also been discussed by them.

For $\bar{k} = 2$, the designs in the dual set-up can be studied directly for **E**-optimality.

2.2 Further E-Optimal Designs

In recent years, there have been some studies on characterization of some *known* designs as **E**-optimal designs in design set-ups $\mathbf{D}(b,v,k)$ where, usually $v \nmid bk$. The following *general* procedure is *apparent* in all the studies made so far.

We start with a *known* **E**-optimal design, say d^*, in a set-up like $\mathbf{D}^*(b^*,v^*,k^*)$ with some combination of values of b^*, v^* and k^*. It is known that for d^*, the minimum positive eigenvalue of the **C**-matrix assumes a known value, say Δ^*. We *modify* $\mathbf{D}^*(b^*,v^*,k^*)$ to the *revised* set-up $\mathbf{D}(b,v,k)$ and change d^* to a corresponding d in $\mathbf{D}(b,v,k)$ in a way that the minimum positive eigenvalue of d can be easily obtained and is given by Δ (say). We now put enough of conditions on the design parameters (b,v,k,b^*,v^*,k^*) such that for any competitor to d in $\mathbf{D}(b,v,k)$, the corresponding minimum positive eigenvalue of its **C**-matrix does *not* exceed Δ. This eventually establishes **E**-optimality of d in $\mathbf{D}(b,v,k)$.

To be precise, then, one needs to start with an **E**-optimal design in a known set-up. *Usually*, a **BBD** or a **GDD** with $\lambda_2 = \lambda_1 + 1$ is taken as the starting design d^* in $\mathbf{D}^*(b^*,v^*,k^*)$. Then a modification to this design d^* is effected by addition (deletion) of m blocks, usually distinct, to (from) d^* or by adding s *extra* treatments to every block of d^*. This results in d in the revised set-up $\mathbf{D}(b,v,k)$ where $b = b^* \pm m$, $v = v^*$, $k = k^*$ or $b = b^*$, $v = v^* + s$, $k = k^* + s$. In each of the cases cited above, the revised value Δ for the derived design d can be easily obtained. Now we put enough of conditions on b^*, v^*, k and m or s such that for no competing design in $\mathbf{D}(b,v,k)$, the minimum positive eigenvalue of its **C**-matrix exceeds Δ. This last part is, of course, highly non-trivial and requires considerable operations with various bounds. Constantine (1981) and Jacroux (1980b, 1983a) have provided some bounds on $kx_{(1)}$ in addition to those we have discussed in subsection 2.1. These are mainly *design-dependent* and/or meant for unequally replicated designs. It must be noted that the revised set-up $\mathbf{D}(b,v,k)$ *usually* leads to the situation $v \nmid bk$ even though for the set-up $\mathbf{D}^*(b^*,v^*,k^*)$ to start with, $v^* | b^*k^*$.

We list below the various bounds that have been suggested in this context.

Constantine (1981) used the 'averaging technique' which states that for any **C**-matrix, the largest and the smallest non-zero eigenvalues of $\bar{\mathbf{C}}$ defined as $\bar{\mathbf{C}} = (\sum_{i=1}^{s} \mathbf{P}_i' \mathbf{C} \mathbf{P}_i)/s$ where $\mathbf{P}_1, \mathbf{P}_2, \ldots, \mathbf{P}_s$ are permutation matrices, are related to those of $\mathbf{C}$ by $x_{(1)} \le \bar{x}_{(1)}$ and $x_{(v-1)} \ge \bar{x}_{(v-1)}$. Averaging separately over *all* permutations of 1 to p and also of $p+1$ to v, we get the following formal representation for $\bar{\mathbf{C}}$.

$$\overline{\mathbf{C}} = \begin{pmatrix} (a+\alpha)\mathbf{I}-\alpha\mathbf{J} & -\beta\mathbf{J} \\ & (b+\gamma)\mathbf{I}-\gamma\mathbf{J} \end{pmatrix}$$

and its non-zero eigenvalues are $a+\alpha$ (repeated $(p-1)$ times), $b+\gamma$ (repeated $v-p-1$ times) and $v\beta$.

Taking $p = 1$ and 2, we get

(a) $kx_{(1)} \leq \frac{vk}{v-1} \min_i C_{ii} \leq \frac{vr_{(1)}(k-1)}{v-1} \leq \frac{vr_0(k-1)}{v-1}$ where $r_{(1)} \leq r_{(2)} \leq \cdots \leq r_{(v)}$ and $r_0 = [bk/v]$.

(b) $kx_{(1)} \leq \frac{k-1}{2}(r_i+r_{i'}) + \lambda_{ii'}$ for all $i \neq i'$.

In particular, if r_i, $r_{i'} \leq r_0$ and $\lambda_{ii'} \leq \lambda_0 = [r_0(k-1)/(v-1)]$, then (b) implies $kx_{(1)} \leq (k-1)r_0 + \lambda_0$.

The above results are also derivable through the technique of principal minor as employed in the proof of Theorem 3.2.4 in subsection 2.1. Jacroux (1983a) made further use of this technique to deduce that

(c) $kx_{(1)} \leq \{\tilde{r}(k-1)-(m-1)z\}v/(v-m)$

where $\tilde{r}$ is the replication number realized by a set of m (say) treatments in a design and $z = \min_{i<i'} \lambda_{ii'}$ over this set of m treatments.

The above expressions provide design-dependent upper bounds to $kx_{(1)}$. There is a similar design-dependent lower bound to $kx_{(1)}$ derived by Jacroux (1980b) using Gershgorin Disc Theorem on a suitable $\mathbf{T}_x$-matrix. This is given by

(d) $kx_{(1)} \geq uv$ where $u = \min_{i<i'} \lambda_{ii'}$.

Below we state the main **E**-optimality results in such contexts. We omit the proofs altogether.

1. **(Constantine (1981))**. Let $0 \leq s < \frac{v}{k}$ be an integer. If s disjoint binary blocks are added to a **BIBD** (b,v,k), the resulting design is **E**-optimal in the class $\mathbf{D}(b+s,v,k)$.
2. **(Constantine (1981))**. Starting with a **GDD** with $\lambda_2 = \lambda_1+1$, if we add $s < \frac{v-m}{k}$ disjoint binary blocks such that each block is a subset of treatments in a group, then the resulting design is **E**-optimal in the class $\mathbf{D}(b+s,v,k)$.
3. **(Constantine (1981), Jacroux (1982))**. Let $v/k^2 \leq s \leq \frac{v}{k}$ be an integer. If s disjoint binary blocks are deleted from a **BIBD** (b,v,k), the resulting design is **E**-optimal in the class $\mathbf{D}(b-s,v,k)$.
4. **(Sathe and Bapat (1985))**. Let $s \leq \frac{v-\sqrt{v}}{v-k}$ be an integer. If *any* s blocks are deleted from a **BIBD** (b,v,k), the resulting design is **E**-optimal in $\mathbf{D}(b-s,v,k)$.
5. **(Jacroux (1983a))**. To a **BIBD** (b,v,k), adding *any* set of $\hat{b}$ blocks each of size k such that $\hat{b}k(k-1) \leq (3v-1)/2$, we get an **E**-optimal design in the class $\mathbf{D}(b+\hat{b},v,k)$.

6. **(Jacroux (1983a))**. To a **GDD** with parameters $m = 2$, $n = v/2$, $\lambda_2 = \lambda_1+2$ or $m = v/2$, $n = 2$, $\lambda_2 = \lambda_1-1$, adding *any* set of $\hat{b}$ blocks each of size k where $\hat{b}k(k-1) \leq (v-1)/2$, we get an **E**-optimal design in the class $\mathbf{D}(b+\hat{b},v,k)$.

Jacroux (1983a) has some further results in the direction of adding some control treatments in every block of a **BIBD** to come up with **E**-optimal designs in the set-up $\mathbf{D}(b,v+s,k+s)$ for $s \geq 1$.

Bagchi (1988) has introduced a new class of designs, called *quotient* designs. An interesting feature of such designs is that these are necessarily non-binary (and non-equireplicate). Under certain conditions on the design parameters, some such designs are shown to be **E**-optimal in the entire class of designs in a given set-up. This contradicts the general belief that non-binary designs are necessarily worse (*inadmissible*) w.r.t. any reasonable optimality criterion. However, in all such cases, a binary **E**-optimal design can also be seen to exist and, hence, it may still be true that the binary designs (for $k<v$) form an *essentially complete* class (This is a decision-theoretic terminology meaning thereby that any competing design outside this class may have at the most the same efficiency as the most efficient design inside this class).

3. Efficiency Factor and A-optimal Designs

Traditionally, an incomplete block design was judged on the basis of the average variance of the estimated paired treatment comparisons. This led to the notion of *efficiency factor* of a block design which was formally defined as follows.

$$\text{Eff. Factor} = E = 2\sigma^2/r\bar{V} \tag{3.3.1}$$

where $2\sigma^2/r$ is the average variance for a Randomized Block Design (**RBD**) (with r blocks of size v each) and $\bar{V}$ is the average variance of the estimated difference between pairs of treatments. Kempthorne (1956) showed that $\bar{V} = 2\sigma^2/H$ where H is the harmonic mean of the non-zero eigenvalues of the **C**-matrix. This leads to the following formal expression for E:

$$E = \frac{H}{\bar{r}} = \frac{v}{bk}\{(\textstyle\sum 1/x_i)/(v-1)\}^{-1} \tag{3.3.2}$$

This also covers the situations where $v \nmid bk$ as $\bar{r} = \dfrac{bk}{v}$ may be regarded as the *average* replication number for an **RBD**. Recalling the definition of **A**-optimality criterion, we observe that the efficiency factor is inversely proportional to the **A**-optimality functional so that an **A**-optimal design maximizes the efficiency factor E. With the idea of achieving **A**-optimal designs, various upper bounds to the efficiency factor have been derived in recent years. In subsection 3.1, we present a detailed description of these bounds in the cases of $v \leq b$ and $v > b$. Most of these bounds are design-dependent in general. But, for some special subclasses of designs, some of them may be made design-independent. Even though these bounds are rarely attained, it is hoped that a study of the relative comparison of various bounds might aid in the search for a *highly efficient* (i.e., *nearly* **A**-optimal) design in a given set-up. It may be pointed out that at this stage very few classes of designs other than those discussed in the previous Chapter are known to be *exactly* **A**-optimal. In subsection 3.2, we present some of them. These are obtained by using the fact that under certain conditions, the **RGDs** do better than the non-**RGDs** w.r.t. a class of optimality criteria including the **A**-optimality criterion. Finally, in subsection 3.3, we introduce the concepts of *Efficiency Balance* (**EB**) and *Variance Balance* (**VB**) and discuss their interrelationship.

3.1 **Upper Bound for the Efficiency Factor**

We shall first deal with the case of $v \leq b$. The other case will be treated at the end. The expression (3.3.2) for the efficiency factor E involves computation of the non-zero eigenvalues x_i's of the **C**-matrix. The expression (3.3.1) suggests that E can be computed with a knowledge of the solution matrix of the normal equations. Following Tocher (1952), we take, for $\epsilon > 0$, a solution matrix of the form

$$\Omega = (\mathbf{C} + \epsilon \mathbf{J}/v)^{-1} \tag{3.3.3}$$

which is p.d..
It can be seen that E now takes the form

$$E = (v-1)/r\{tr(\Omega) - \epsilon^{-1}\}. \tag{3.3.4}$$

Below we will work with both (3.3.2) and (3.3.4). The expression (3.3.2) suggests that an upper bound to E may be obtained by minimizing $\sum 1/x_i$. Using the fact that $(\sum 1/x_i)(\sum x_i) \geq (v-1)^2$ and that $\sum x_i = tr(\mathbf{C})$ is a maximum when the design is binary, it follows that the *first* upper bound to E is given by

$$E_I = \frac{v(k-1)}{k(v-1)} + \frac{v[k/v]\{v[k/v] - 2k + v\}}{k^2(v-1)}. \tag{3.3.5}$$

It may be checked that $E_I \leq 1$ with equality iff $k = 0 \ (mod\ v)$ in which case a design consisting of k/v copies of an **RBD** attains this bound. When $k < v$, E_I reduces to $(1-k^{-1})/(1-v^{-1})$ which is attained by a **BIBD** with parameters (b,v,k). Otherwise, it is reached by a **BBD**. (In this context, see Remark 2.2.1 in Chapter Two). To get to other tighter bounds on the efficiency factor E, we will use the following notations.

For a given design, we write (as in section 3, Chapter Two)

$$A = tr(\mathbf{C}), \ B = tr(\mathbf{C}^2), \ P^2 = B - A^2/(v-1)$$

and let, further, $D = tr(\mathbf{C}^3)$.

It may be noted that the bound (3.3.5) is achieved only when $P = 0$ and A is a maximum. In case $P>0$, we may refer to the analysis of Conniffe and Stone (1974) as used earlier in Chapter Two. This gives an upper bound to H in (3.3.2) as

$$(v-1)[\{(v-2)/(A-\theta P)\} + \{A+\theta(v-2)P\}^{-1}]^{-1} \tag{3.3.6}$$

with $\theta = \{(v-1)/(v-2)\}^{1/2}$. This bound is attained when the eigenvalues assume only two distinct values viz., $(A-\theta P)/(v-1)$ with multiplicity $(v-2)$ and $(A + \theta(v-2)P)/(v-1)$ with multiplicity one.

Jarrett (1977) gave an alternative expression to (3.3.6) in the form

$$\{A/(v-1)\} - \{P^2/(v-1)\}[\{A/(v-1)\} + \{(v-3)P/(v-1)^{1/2}(v-2)^{1/2}\}]^{-1}. \tag{3.3.7}$$

Clearly this bound is design-dependent. In view of the preceding analysis on the *nature* of optimal designs so far presented, it seems reasonable to restrict our attention to the subclass of designs which maximize A and subsequently, also minimize P. Further, for simplicity, we will henceforth work with the case of $k<v$ and $bk = vr$, r an integer. Thus, essentially, we will talk of bounds for binary equireplicate designs having the concurrences $\lambda_{ii'}$'s differ at most by unity. In other words, we will deal with the **RGDs**.

Following Jarrett (1977), we write

$$r(k-1) = (\delta+\alpha)(v-1),\ \ \delta\,(\text{an integer})\geq 1,\ \ 0\leq\alpha<1.$$

It can be easily seen that for such designs,

$$A = vr(k-1)k^{-1} \text{ and } P^2 = v(v-1)\alpha(1-\alpha)k^{-2}. \tag{3.3.8}$$

Dean and Lewis (1984) substituted the above expressions for A and P^2 in (3.3.7) to deduce that the *second* upper bound to E can be written as

$$E_{II} = (E_I-S)\{E_I+(v-2)S\}\{E_I+(v-3)S\}^{-1} \tag{3.3.9}$$

where $S^2 = P^2/(v-1)(v-2)$. Clearly $E_{II}\leq E_I$ with "=" iff $S = 0$ i.e., $P = 0$. Further E_{II} is reached iff the design is a **GDD** with $m = 2$ and $\lambda_2 = \lambda_1+1$. (Vide subsection 3.3 in Chapter Two).

Pearce (1968) and Williams and Patterson (1977) derived other bounds to the efficiency factor based on an analysis of its representation in (3.3.4). Towards this, first observe that for the **RGDs** with $k<v<b$ and $bk = vr$, r an integer, Ω in (3.3.3) yields, for $\epsilon>0$,

$$\Omega^{-1} = \mathbf{C}+\epsilon\mathbf{J}/v = r(\mathbf{I}_v-\mathbf{G}_\epsilon)$$

where $\mathbf{G}_\epsilon = r^{-1}k^{-1}\mathbf{NN}'-\epsilon\mathbf{J}/vr$ is symmetric with eigenvalues $1-\epsilon/r$ and $1-x_i/r$, $1\leq i\leq v-1$. Note that x_i's are the non-zero eigenvalues of $\mathbf{C}$ so that $x_i\leq r$ for $1\leq i\leq v-1$. Hence, for *any* choice of $\epsilon<2r$, the eigenvalues of $\mathbf{G}_\epsilon$ are less than unity in absolute value. This justifies the expansion $(\mathbf{I}_v-\mathbf{G}_\epsilon)^{-1} = \sum_{j=0}^{\infty}\mathbf{G}_\epsilon^j$ with $\mathbf{G}_\epsilon^0 = \mathbf{I}_v$. By taking $\epsilon\leq r$, we can actually make $\mathbf{G}_\epsilon$ an nnd matrix and, hence, $(\mathbf{I}_v-\mathbf{G}_\epsilon)^{-1} \geq \sum_{j=0}^{m}\mathbf{G}_\epsilon^j$ for every $m\geq 1$. In particular, taking $\epsilon = r$ and $m = 2$, Pearce (1968) deduced that

$$E \leq (v-1)/r\{r^{-1}tr(\mathbf{I}+\mathbf{G}_r+\mathbf{G}_r^2)-r^{-1}\} = (v-1)/\{(v-1)+tr(\mathbf{G}_r)+tr(\mathbf{G}_r^2)\}.$$

However, $tr(\mathbf{G}_r) = (vk^{-1}-1)$ and $tr(\mathbf{G}_r^2) = (vk^2-1) + v(v-1)\alpha(1-\alpha)r^{-2}k^{-2} + \{v/(v-1)\}\{(k-1)/k\}^2$ for the **RGDs** with $k<v<b$, $bk = vr$ (r an integer) and $r(k-1) = (\delta+\alpha)(v-1)$, $\delta\geq 1$ (an integer) and $0\leq\alpha<1$. Thus, finally, we get (as in Jarrett (1977) and Dean and Lewis (1984)),

$$E \leq (v-1)/[(v-1)+(vk^{-1}-1)+(vk^{-2}-1)+v(v-1)\alpha(1-\alpha)r^{-2}k^{-2}+v(v-1)^{-2}(k-1)^2k^{-2}]$$

which, on simplification, gives the *third* bound

$$E_{III} = \{3-E_I+E_I^2+(v-2)S^2\}^{-1}. \tag{3.3.10}$$

Williams and Patterson (1977) took one more term in the expansion of $(\mathbf{I}_v-\mathbf{G}_r)^{-1}$ and got, formally,

$$E \leq (v-1)/\{(v-1)+tr(\mathbf{G}_r)+tr(\mathbf{G}_r^2)+tr(\mathbf{G}_r^3)\}.$$

Clearly, the task ahead is to get a convenient expression for $tr(\mathbf{G}_r^3)$. Towards this, note that $\mathbf{G}_r = \mathbf{G}_r^*-b\mathbf{J}/v$ where $b = 1 - \frac{\delta v}{rk}$ and $\mathbf{G}_r^* = r^{-1}k^{-1}\{(r-\delta)\mathbf{I}+\mathbf{W}\}$. Here $\mathbf{W}$ is a $(v\times v)$ symmetric matrix of 0's and 1's essentially representing the deviations of $\mathbf{NN}'$ from a completely symmetric form. We see easily that $tr(\mathbf{W}) = 0$ and $tr(\mathbf{W}^2) = v(v-1)\alpha$. We next observe that $\mathbf{G}_r^*\mathbf{J} = b\mathbf{J}$ and, hence,

$$\begin{aligned} tr(\mathbf{G}_r^3) &= tr(\mathbf{G}_r^{*3})-b^3 = r^{-3}k^{-3}tr\{(r-\delta)\mathbf{I}+\mathbf{W}\}^3-b^3 \\ &= r^{-3}k^{-3}\{v(r-\delta)^3+3(r-\delta)v(v-1)\alpha+tr(\mathbf{W}^3)\} - (1-\frac{\delta v}{rk})^3. \end{aligned}$$

Thus, for an **RGD**, even though the bound E_{III} is design-independent, the one involving $tr(\mathbf{G}_r^3)$ is not so. To overcome this difficulty, Williams and Patterson (1977) used some combinatorial arguments and obtained a non-trivial lower bound to $tr(\mathbf{W}^3)$. We omit the details and simply state that $tr(\mathbf{W}^3) \geq 0$ if $\alpha(v-1) \leq v/2$ and $\geq v(v-1)\alpha\{2\alpha(v-1)-v\}$ otherwise. This led them to finally come up with a *fourth* upper bound for E. Dean and Lewis (1984) have put it in the following form:

$$E_{IV} = E_I/[1+(v-2)S^2E_I^{-3}r^{-1}k^{-1}\{rkE_I-z+\alpha(v+1)\}] \qquad (3.3.11)$$

where $z = \alpha(v-2)/(1-\alpha)$ for $\alpha(v-1)<v/2$ and $= v$ otherwise.

We refer to (3.3.2) once more and mention yet another bound discussed by Jarrett (1983) using an inequality of independent interest. Fitzpatrick and Jarrett (1983) have observed that the harmonic mean (H) of a set of positive real numbers $x_1, \ldots, x_n$ obeys the inequality $H \leq \mu_1' - \mu_2^2/\{\mu_3+\mu_1'\mu_2\}$ where μ_1', μ_2 and μ_3 refer to their mean and the second and third central moments respectively. Here "=" holds iff x_i's comprise of *two* distinct values only. Using this, Jarrett (1983) developed bounds to E in (3.3.2) for binary, equireplicate designs with two distinct concurrences, and, in particular, for the 2-associate **PBIBDs** (in which case the bound is attained). These bounds are, as such, design-dependent. Specifically, for the **RGDs**, one may again use the lower bound to $tr(\mathbf{W}^3)$ discussed above to come up with the following design-independent *fifth* upper bound to the efficiency factor as given in Dean and Lewis (1984).

$$E_V = \frac{\{E_I^2-(v-2)S^2+E_Ir^{-1}k^{-1}(z-\alpha(v+1))\}}{\{E_I+r^{-1}k^{-1}(z-\alpha(v+1))\}} \qquad (3.3.12)$$

where z is as given immediately after (3.3.11).

These are some bounds for the efficiency factor of designs in the subclass of the **RGDs** for given (b,v,k) with $k<v<b$ and $bk = vr$, r an integer. These bounds have been compared algebraically and numerically for various combinations of values of (b,v,k). It has been concluded that

(i) $E_I \geq E_{II}, E_{IV}$;

(ii) $E_V \leq E_{II}, E_{IV}$;

(iii) E_{III} behaves erratically but most often it even exceeds E_I and hence is *not* a useful bound;

(iv) there is no definite ordering between E_{II} and E_{IV}.

Designs attaining bounds E_I and E_{II} have been characterized earlier. These are the **BBDs** or the **GDDs** with $m = 2$ and $\lambda_2 = \lambda_1+1$ respectively. Even though the above analysis suggests that these designs are **A**-optimal in the subclass of the **RGDs**, it is known that they are optimal in the entire class $\mathbf{D}(b,v,k)$ w.r.t. the generalized optimality criteria. We note that when E_I attains, $E_I = E_{II} = E_{IV} = E_V < E_{III}$ and when E_{II} attains, $E_{II} = E_V < E_I$, E_{III}, E_{IV}. It is difficult to characterize designs attaining the bounds E_{III} and E_{IV}. This is mainly because these bounds are based on the first few terms in an infinite series expansion of $(\mathbf{I}_v - \mathbf{G}_r)^{-1}$. As regards E_V, it is easy to see that this will be reached when the design is a two-associate **PBIBD** with $|\lambda_1 - \lambda_2| = 1$ and $tr(\mathbf{W}^3)$ attains its relevant lower bound. Interestingly enough, we observe that at least for two classes of designs viz., (i) **GDDs** with $\lambda_2 = \lambda_1 + 1$, and (ii) **GDDs** with $n = 2$, $\lambda_1 = \lambda_2 + 1$, $tr(\mathbf{W}^3)$ assumes the values $v(v-n)(v-2n)$ and 0 respectively which are indeed the relevant lower bounds. Thus these designs are **A**-optimal within the subclass of the **RGDs**. (It may be recalled that such designs are known to be **E**-optimal in the entire class of designs in $\mathbf{D}(b,v,k)$). Clearly, some further tools are needed to establish the property of **A**-optimality of these designs in the entire class. This will be taken up in the next subsection.

As regards the case of $v>b$ (which also covers the study of dual designs), we first note that the efficiency factor E in such a set-up is given by

$$E = (v-1)[\{(b-1)/\bar{E}\} + (v-b)]^{-1} \tag{3.3.13}$$

where $\bar{E}$ refers to the efficiency factor in the dual set-up with $\bar{b} = v > \bar{v} = b$ and $\bar{k} = r = bk/v$. Thus if $v>b$ and we confine to designs whose duals are **RGDs**, we first get an upper bound to $\bar{E}$ using any of the above procedures and then apply (3.3.13) to get a corresponding bound to E. The bound in (3.3.13) uses the fact that when $v>b$, at least $(v-b)$ eigenvalues of the **C**-matrix must equal r. On the other hand, for computation of E_I and E_{II}, we did *not* require $b>v$ and, therefore, for a given design with $b<v$, we can directly compute the efficiency bounds E_I and E_{II}. However, the bounds derived by using (3.3.13) are expected to be sharper.

3.2 A-optimality of Certain RGDs

In order to justify the search for optimal designs to the subclass of the **RGDs**, it is enough to get hold of an **RGD** which does better than *any* non-**RGD**. Once we succeed in this, then it is clear that the best **RGD** is **A**-optimal in the entire class.

Jacroux (1985) developed certain upper bounds to $\sum_1^{v-1} f(kx_i)$ for *any* non-**RGD** in a given design set-up w.r.t. an arbitrary but fixed optimality functional f satisfying the usual properties $f'<0$, $f''>0$, $f'''<0$ as listed in subsection 3.3 in Chapter One. Below we discuss his results only for the case of $v \mid bk$ and $k<v$. Let, as usual, A and B denote $tr(\mathbf{C})$ and $tr(\mathbf{C}^2)$ respectively for an **RGD**. Now if a design is non-binary, then $tr(\mathbf{C})$ is reduced by at least $2/k$. On the other hand, if a design is binary but a non-**RGD**, the value of $tr(\mathbf{C})$ remains the same as A while $tr(\mathbf{C}^2)$ gets increased by at least $4/k^2$ when $k \geq 3$, and by $1/2$ when $k=2$. Write $A^* = A - 2/k$ and $B^* = B + \min(1/2, 4/k^2)$. Then, in the former case, if m_1^* is the bound to $kx_{(1)}$ for a non-binary design, it is shown that

$$\sum_{1}^{v-1} f(kx_{(i)}) \geq f(m_1^*) + (v-2)f(\frac{kA^* - m_1^*}{v-2}). \tag{3.3.14}$$

In the latter case, again, if m_1 is the bound to $kx_{(1)}$ for a binary non-**RGD**, it is shown that

$$\sum_{1}^{v-1} f(kx_{(i)}) \geq f(m_1) + \{(v-3)f(m_2) + f(m_3)\} \tag{3.3.15}$$

where the bracketted quantity $\{\cdots\}$ in the above is derived by using Conniffe and Stone-type argument to $\sum_{2}^{v-2} f(kx_{(i)})$ with $\sum_{2}^{v-2}(kx_{(i)}) = Ak - m_1$ and $\sum_{2}^{v-2}(kx_{(i)})^2 = k^2B^2 - m_1^2$. If now for an **RGD**, $\sum_{1}^{v-1} f(kx_{(i)})$ is less than both the bounds in (3.3.14) and (3.3.15), then the *best* **RGD** is optimal w.r.t. the particular optimality function f. The bounds m_1^* and m_1 used by Jacroux (1985) in this context are the ones we have already studied in subsection 2.1. Specifically, these are

$$m_1^* = \max\{\frac{(r(k-1)-2)v}{v-1}, \frac{(r-1)(k-1)v}{v-1}\} \tag{3.3.16}$$

and

$$m_1 = \max\{h_{\delta+2}, g_{\delta-1}, \frac{(r-1)(k-1)v}{v-1}\}. \tag{3.3.17}$$

We refer to subsection 2.1 for the details.

When an **RGD** satisfying the above requirements is *unique* (in the sense of producing a unique **C**-matrix), it is indeed optimal w.r.t. the specific ψ_f-optimality criterion.

The above discussion gives a general approach for establishing optimality of the **RGD**s. In particular, taking $f(x) = 1/x$, we may establish **A**-optimality of certain **RGD**s in the entire class. John and Mitchell (1977) made an extensive search of the *best* (**A**-optimal) **RGD**s and displayed them for all combinations of (b,v,k) with $v \leq 12$, $r \leq 10$ and $v \leq b$. Jacroux (1985) has verified that except for the case of $b = v = 8$, $r = k = 3$, *all* other designs listed by John and Mitchell (1977) for $k \geq 3$, $v \leq 8$, satisfy the sufficient condition described above and, hence, are **A**-optimal in the entire class $\mathbf{D}(b,v,k)$. Our computations indicate that even for the case of $b = v = 8$, $r = k = 3$, the **RGD** given by John and Mitchell (1977) *does* satisfy the required conditions and, hence, it is **A**-optimal. Jacroux (1985) has also listed some situations where an **RGD** is unique. The two classes of **GDD**s described in the previous section viz., the **GDD**s with $\lambda_2 = \lambda_1 + 1$ and those with $\lambda_1 = \lambda_2 + 1$, $n = 2$ are **A**-optimal within the respective sub-classes of **RGD**s for given (b,v,k). If for such a design, the eigenvalues satisfy the relevant sufficient condition relating to

A-optimality, then the design is A-optimal in the entire class.

When bk is *not* divisible by v, an **RGD** is *not* available. In that case, an **SRGD** i.e., a binary design with $|r_i - r_{i'}| \leq 1$ and $|\lambda_{ii'} - \lambda_{ss'}| \leq 1$ may exist. Jacroux (1985) has derived sufficient conditions for the A-optimality of **SRGDs** in a manner similar to the one described above for the **RGDs**. We mention in passing that Cheng and Wu (1981) have provided a table showing A- and D-efficiencies of **SRGDs** which they have termed *Nearly Balanced Incomplete Block Designs*. Their table covers the cases of $v = 5$ and 6.

3.3 Concept of Efficiency Balance

The notion of canonical or basic contrasts is sometimes used in the study of the efficiency of the designs. This may be described as follows.

We consider a block design in which the i-th treatment is replicated r_i times; $i = 1,2, \ldots, v$. We also consider a *Completely Randomized Design* (**CRD**) with the same replication numbers for the treatments. We now seek a complete set of $(v-1)$ treatment contrasts which possess uncorrelated estimates w.r.t. each of the above two designs. Towards this we first obtain an orthogonal set of eigenvectors for $\mathbf{A} = r^{-\delta/2}\mathbf{C}\mathbf{r}^{-\delta/2} = \mathbf{I} - \mathbf{r}^{-\delta/2}(\mathbf{NN}'/k)\mathbf{r}^{-\delta/2}$ where $\mathbf{r}^{c\delta} = diag(r_1^c, \ldots, r_v^c)$, c being a scalar. Call these $\mathbf{u}_1, \mathbf{u}_2, \ldots, \mathbf{u}_v$ where in view of the fact that $\mathbf{C1} = 0$ we may take $\mathbf{u}_1 = \mathbf{r}^{\delta/2}\mathbf{1}$. It is easy to see that if $\mathbf{s}_i = \mathbf{r}^{\delta/2}\mathbf{u}_i$, $i = 2,3, \ldots, v$, $\mathbf{s}_i'\tau$'s are treatment contrasts with the required property where τ is the vector of treatment effects. For $\mathbf{s}_i'\tau$, the ratio of the variance of $\mathbf{s}_i'\hat{\tau}$ for **CRD** to that for the block design is ϵ_i where ϵ_i is the eigenvalue of $\mathbf{A}$ associated with $\mathbf{u}_i$. Thus, ϵ_i may be termed the efficiency of the block design for the contrast $\mathbf{s}_i'\tau$. These contrasts are called *basic* or *canonical* contrasts and the corresponding ϵ_i's are called *canonical efficiencies*. It has been observed that for some designs the analysis in terms of the basic contrasts is very simple.

It is clear that $0 < \epsilon_i \leq 1$ for any connected design. The harmonic mean of the ϵ_i's may be called the *canonical efficiency factor* of the design.

We note that when $r_1 = \cdots = r_v$, this canonical efficiency factor coincides with the efficiency factor E defined in (3.3.2). However, when the replication numbers are unequal, the canonical efficiency factor does not seem to have any obvious statistical interpretation similar to that for E.

A design is called *Efficiency Balanced* (**EB**) (Puri and Nigam (1975)) if the ϵ_i's are all equal to ϵ (say). It is easy to see that this holds iff $\mathbf{A} = \epsilon(\mathbf{I} - \mathbf{r}^{\delta/2}\mathbf{J}\mathbf{r}^{\delta/2}/bk)$ i.e. $\mathbf{C} = \epsilon(\mathbf{r}^{\delta} - \mathbf{r}^{\delta}\mathbf{J}\mathbf{r}^{\delta}/bk)$. This gives $\mathbf{NN}' = (k-\epsilon)\mathbf{r}^{\delta} + \epsilon\mathbf{r}^{\delta}\mathbf{J}\mathbf{r}^{\delta}/b$. Hence if a design is efficiency balanced, $\lambda_{ii'}$ is proportional to $r_i r_{i'}$. It is easy to see that the converse is also true. For an **EB** design, any treatment contrast is a canonical contrast with the same efficiency. In contrast to this, a design for which all normalized contrasts possess the same variance is called *variance balanced* (**VB**) design. For such a design, the non-zero eigenvalues of the C-matrix are all equal and consequently the C-matrix assumes a completely symmetric form.

In the set-up of $\mathbf{D}(b,v,k)$, some interesting relationships between **VB** and **EB** are known. For example, a **VB** design is necessarily equireplicate and, hence, is also **EB**. Further, an **EB** design which is binary and/or equi-replicate, is necessarily **VB**. Thus non-trivial examples of **EB** designs are to be found among designs which are non-binary *and* non-equireplicate. We mention in passing that the balanced ternary designs of Tocher (1952) are **VB** and equireplicate and, hence, **EB**.

Recently there has been a substantial amount of work on the combinatorial and constructional aspects of **EB** and/or **VB** designs. Since this is *not* directly related to optimality aspects of designs, we will *not* pursue this topic further. Some useful references are Hedayat and Federer (1974), Kageyama (1980), Puri and Nigam (1977), Pearce et al. (1974) and James and Wilkinson (1971).

4. MV-optimal Designs

In Chapter One (subsection 3.4), we described one optimality criterion due to Takeuchi (1961) which has been called **MV**-optimality criterion by Jacroux (1983b). We recall that a design in the class $\mathbf{D}(b,v,k)$ is said to be **MV**-optimal if it minimizes the maximum variance for a paired treatment contrast among all designs in $\mathbf{D}(b,v,k)$. Thus, **MV**-optimality is somewhat different from **E**-optimality in which the comparison refers to *all* treatment contrasts. It is also different from **A**-optimality which seeks to relate to the average variance for *all* paired treatment contrasts. However, unlike the **A**-, **D**- and **E**-optimality criteria, this criterion is *not* exclusively a function of the eigenvalues and, as such, it needs a separate treatment. Since the **BBDs** are universally optimal, they are also **MV**-optimal. Some other classes of asymmetrical designs are also known to be **MV**-optimal. We present below the optimality results due to Takeuchi (1961) and Jacroux (1983b).

The general approach to the search for an **MV**-optimal design may be sketched as follows. Since such an optimal design seeks to minimize the maximum variance, it would be useful to obtain a lower bound to the variance of a paired treatment contrast for an arbitrary design in $\mathbf{D}(b,v,k)$. Towards this, we develop a lower bound to the variance $V_{ii'}$ of the estimate of the treatment contrast involving the treatment pair (i,i'). Referring to subsection 3.1, we note that the matrix Ω in (3.3.3) serves as the (pseudo) variance-covariance matrix of the estimates of treatment effects. Thus a formal expression for the variance of the above contrast is given by $V_{ii'} = \sigma^2(\mathbf{e}_{ii'}'\Omega\mathbf{e}_{ii'})$ where $\mathbf{e}_{ii'}$ is a $(v\times1)$ column vector with 1 in the ith position and -1 in the i'th position. Denote by $\Omega_{ii'}$ the 2×2 principal minor of Ω involving the rows i, i' and the corresponding columns. Thus

$$\mathbf{e}_{ii'}'\Omega\mathbf{e}_{ii'} = (1\ -1)\Omega_{ii'}(1\ -1)'. \tag{3.4.1}$$

Recalling the expression for Ω in (3.3.3) viz., $\Omega = (\mathbf{C}+\epsilon\mathbf{J}/v)^{-1}$ with $\epsilon>0$, we note that

$$\Omega_{ii'} \geq \begin{pmatrix} C_{ii}+\epsilon/v & C_{ii'}+\epsilon/v \\ & C_{i'i'}+\epsilon/v \end{pmatrix}^{-1}.$$

We now choose $\epsilon = v\lambda_{ii'}/k = -vC_{ii'}$ so that

$$\sigma^{-2}V_{ii'} \geq (C_{ii}-C_{ii'})^{-1} + (C_{i'i'}-C_{ii'})^{-1}.$$

Using the fact that $kC_{ii} \leq r_i(k-1)$ for all i, we further deduce that

$$\sigma^{-2}V_{ii'} \geq \frac{k}{r_i(k-1)+\lambda_{ii'}} + \frac{k}{r_{i'}(k-1)+\lambda_{ii'}}$$

and this finally yields

$$V_{ii'} \geq \sigma^2 \frac{4k}{\{(r_i+r_{i'})(k-1)+2\lambda_{ii'}\}}. \tag{3.4.2}$$

Thus if for a competing design the maximum variance of paired treatment contrasts is less than $\sigma^2 4k/\{(r_i+r_{i'})(k-1)+2\lambda_{ii'}\}$ for *some* treatment pair (ii'), $i \neq i'$, the competing design is better w.r.t. the **MV**-optimality criterion.

Takeuchi (1961) essentially used this idea and came up with **MV**-optimality of the **GDDs** with $\lambda_2 = \lambda_1+1$. The precise result in stated and proved below. We assume $bk = vr$, r an integer and $k<v$.

Theorem 3.4.1. If, for given (b,v,k), a **GDD** having parameters m, n, λ_1, λ_2 with $\lambda_2 = \lambda_1+1$ exists, it is **MV**-optimal in $\mathbf{D}(b,v,k)$.

Proof. It can be easily checked that for a **GDD** with $\lambda_2 = \lambda_1+1$, the maximum variance for a paired treatment contrast is attained for any within group contrast and it is given by

$$2k\sigma^2/\{r(k-1)+\lambda_1\}. \tag{3.4.3}$$

Suppose now that a competing design is equireplicate. As $r(k-1) = \lambda_1(v-1)+(v-n) = \sum_{i'=2}^{v}\lambda_{1i'}$, it is clear that for the competing design, not all $\lambda_{1i'}$ can exceed λ_1. So we take $\lambda_{1i'} \leq \lambda_1$. In (3.4.2), we then take $i = 1$ and get for such a competitor $V_{1i'} \geq 4k\sigma^2/\{2r(k-1)+2\lambda_1\}$ which is (3.4.3). Again if a competing design is non-equireplicate, $r_1 \leq r-1$. We claim that for some i', $(r_1+r_{i'})(k-1)+2\lambda_{1i'} \leq 2r(k-1)+\lambda_1$. Observe that $\sum_{i'=2}^{v}\{(r_1+r_{i'})(k-1)+2\lambda_{1i'}\} = (v-1)(k-1)r_1$ $+(k-1)(vr-r_1)+2r_1(k-1) = (k-1)v(r+r_1) \leq v(k-1)(2r-1) = 2r(v-1)(k-1)+(2r-v)(k-1)$ $= 2r(k-1)(v-1)+2\lambda_1(v-1)+(3v-vk-2n)$.

Clearly, $v(3-k)-2n$ is negative for all $k \geq 3$, and is less than $v-1$ for $k = 2$. As $\lambda_{1i'}$'s are integers, the minimum value of $(r_1+r_{i'})(k-1)+2\lambda_{1i'}$ certainly cannot exceed $2r(k-1)+2\lambda_1$. This establishes our claim.

Thus for any competing design, $V_{ii'}$ is at least equal to the expression in (3.4.3) for some (i,i'), $i \neq i'$. This shows that the **GDD** is **MV**-optimal in $\mathbf{D}(b,v,k)$.

Remark 3.4.1. It also follows from the above that for $k \geq 3$, the **GDDs** are *uniquely* **MV**-optimal. This result easily extends to the case when $k > v$ and a generalized **GDD** exists.

Jacroux (1983b) used a similar technique to prove the following optimality results. We omit the proof.

Recall the representation $r(k-1) = (\delta+\alpha)(v-1)$, δ an integer, $0 \leq \alpha < 1$. Suppose $2(\delta-1) \geq [(2r-v)(k-1)/(v-1)]$, the integral part of $\{(2r-v)(k-1)(v-1)^{-1}\}$. Consider the following two classes of **GDDs**:

(I) $v = mn$, $m = v/2$, $n = 2$, $\lambda_2 = \lambda_1-1$

(II) $v = mn$, $m = 2$, $n = v/2$, $\lambda_2 = \lambda_1+2$.

If a design of type (I) exists, it is **MV**-optimal. If it does *not* exist but a design of type (II) exists, then the latter one is **MV**-optimal. In this case an **IIGBBD** of the type mentioned in Remark 3.2.6 is also **MV**-optimal.

Jacroux (1983b) also succeeded in establishing **MV**-optimality of some classes of unequally replicated designs in settings where $v \nmid bk$. Some such designs are obtained by adding one or more blocks to certain **BIBDs**. Further, some designs given by Adhikary (1965) are also shown to be **MV**-optimal.

We conclude this section by establishing **MV**-optimality of the minimal covering designs studied by Roy and Shah (1984). They had established optimality of such designs w.r.t. the generalized optimality criteria and had conjectured that such designs are also **MV**-optimal. This is indeed true and can be established without much difficulty following the above approach. Below we indicate the main steps towards this. We refer to subsection 3.4 of Chapter Two for the details.

(i) For such a design, the maximum variance for a paired treatment contrast is $6\sigma^2/v$.

(ii) For any competing design, using (3.4.2), $V_{ii'} \geq 6\sigma^2/\{r_i+r_{i'}+\lambda_{ii'}\}$ for every pair (i,i'), $i \neq i'$.

(iii) We need to show $r_i+r_{i'}+\lambda_{ii'} \leq v = 6t+5$ for *some* (i,i'), $i \neq i'$.

(iv) This follows *by contradiction* which is arrived at by choosing an $r_i \leq r_0 = [bk/v] = 3t+2$ and adding over *all* $i'(\neq i)$.

5. D-optimal Designs

As explained previously in Chapter One, the criterion of **D**-optimality relates to the consideration of minimizing the volume of the concentration ellipsoid while providing an ellipsoidal confidence region for any set of $(v-1)$ linearly independent treatment contrasts. Unlike the **A**-optimality criterion which involves *all* paired treatment contrasts, the **D**-optimality criterion can be viewed as one involving only $(v-1)$ of them which are, of course, taken to be linearly independent. Again it may be noted that while the **E**- and **MV**-optimality criteria have a *minimax* character, the **A**- and **D**-optimality criteria convey a sense of *averaging*. Recall that the **D**-optimality criterion can be expressed as $\psi_f(x_1, \ldots, x_{v-1}) = -\sum_1^{v-1} \log x_i$ and the objective is to minimize $\psi_f(.)$ or, equivalently, to maximize $\prod_1^{v-1} x_i$. Here x_i's denote, as usual, the non-zero eigenvalues of the **C**-matrix.

In Chapter Two, we established universal optimality of the **BBDs** (section 2) and optimality of the **GDDs** with $m = 2$, $n = v/2$, $\lambda_2 = \lambda_1 + 1$ as also of a class of minimal covering designs w.r.t. the generalized optimality criteria (subsection 3.3). Therefore these designs are **D**-optimal in the class $\mathbf{D}(b,v,k)$. Further, Cheng (1979) has also provided **D**-optimal designs for the case of $v = 4$ and these have been stated earlier in subsection 3.5 of Chapter Two. Apart from these results, *not much* is known regarding **D**-optimality of designs in the design setting $\mathbf{D}(b,v,k)$. Gaffke (1982) discussed the cases of $v \leq 6$ using graph-theoretic methods and established **D**-optimality of certain designs. Jacroux (1984) established **D**-optimality of the GDDs with $m = v/2$, $n = 2$, $\lambda_2 = \lambda_1 \pm 1$ for any $k \geq 3$. Further to this, Jacroux (1985) also provided some results on the reduction of the search for **D**-optimal designs to the subclass of the **RGDs**. This is done in a manner very similar to the one used in the context of **A**-optimality which is described in section 3.

We first present a precise statement of the result in Jacroux (1984) and give a sketch of the proof.

Theorem 3.5.1. For given (b,v,k) with $k \geq 3$, if there exists a **GDD** with parameters $m = v/2$, $n = 2$, $\lambda_2 = \lambda_1 \pm 1$, then it is **D**-optimal in the entire class $\mathbf{D}(b,v,k)$.

Proof. The task is clearly to maximize the product of the non-zero eigenvalues of the **C**-matrices among all competing designs in $\mathbf{D}(b,v,k)$. Define $\mathbf{T}_\epsilon = k\mathbf{C} + \epsilon\mathbf{J}/v$ so that $|\mathbf{T}_\epsilon| = \{\prod_1^{v-1} kx_{(i)}\}\epsilon$ and, hence, the problem can be viewed as that of maximizing $|\mathbf{T}_\epsilon|$ for some $\epsilon > 0$. It can be easily checked that for a **GDD** with parameters $m = v/2$, $n = 2$, $\lambda_2 = \lambda_1 \pm 1$, the non-zero eigenvalues of the **C**-matrix are $v\lambda_2/k$ with multiplicity $(m-1)$ and $\{r(k-1)+\lambda_1\}/k$ with multiplicity m $(=v/2)$. Taking $\epsilon = v\lambda_2$, we see that for the above **GDD**, $|\mathbf{T}| = (v\lambda_2)^m\{r(k-1)+\lambda_1\}^m$. Note, however, that $r(k-1) = \lambda_1(n-1)+\lambda_2(v-n)$ so that $v\lambda_2 = \{r(k-1)+\lambda_2\} + (\lambda_2-\lambda_1)$ and $r(k-1)+\lambda_1 = \{r(k-1)+\lambda_2\} - (\lambda_2-\lambda_1)$. Hence, for the above **GDD** with $\lambda_2 = \lambda_1 \pm 1$, $|\mathbf{T}|$ takes the value

$$|\mathbf{T}| = (t^2-1)^{v/2}, \quad t = r(k-1)+\lambda_2. \tag{3.5.1}$$

As to the competing designs in $\mathbf{D}(b,v,k)$, we may classify them as (i) binary and (ii) non-binary. For designs belonging to each of these categories, we need to show that the quantity $|\mathbf{T}| = |k\mathbf{C}+\lambda_2\mathbf{J}|$ is less than or equal to $(t^2-1)^{v/2}$ with t as in (3.5.1). Below we dispose of the case of non-binary designs which is easy to handle. The case of binary designs is more involved and we refer to Jacroux (1984) for the proof.

Suppose a competing design is non-binary. Then it is either non-equireplicate or equireplicate with some $n_{ij} \geq 2$. We denote by t_{ii} the ith diagonal element of the corresponding $\mathbf{T}$-matrix. Then, in the first case, $t_{ii} \leq (r-1)(k-1)+\lambda_2$ assuming $r_i \leq r-1$. Also in the second case, $t_{ii} \leq r(k-1)+\lambda_2-2$. In either case, $t_{ii} \leq t-2$ where $t = r(k-1)+\lambda_2$. Now for a competing design, $|\mathbf{T}| \leq t_{ii} \prod_{j(\neq i)} t_{jj}$. Clearly, $\sum_{j(\neq i)} t_{jj}+t_{ii} \leq vt-2$ as vt is the maximum of $tr(\mathbf{T})$ which is attained only for binary designs. It is *not* difficult to verify that the maximum of $t_{ii} \prod_{j(\neq i)} t_{jj}$ subject to $t_{ii} \leq t-2$ and $\sum_1^v t_{jj} \leq vt-2$ is attained at $t_{ii} = t-2$ and $t_{jj} = t$ for all $j(\neq i)$. As a result, for *any* competing design, $|T| \leq (t-2)t^{v-1} = (t^2)^{v/2-1}(t^2-2t)$. As $(v/2-1)t^2 + (t^2-2t) = (v/2)(t^2 - 4t/v)$, we deduce further that $|\mathbf{T}| \leq (t^2 - 4t/v)^{v/2}$. It is now straightforward to verify that $4t \geq v$ in the present set-up with $t = r(k-1)+\lambda_2$ so that finally, $|\mathbf{T}| \leq (t^2-1)^{v/2}$ which is (3.5.1). Hence the result.

Remark 3.5.1. As seen in section 2, these designs are known to be $\mathbf{E}$-optimal in the entire class. Moreover, in section 3 it has been shown that such designs are also $\mathbf{A}$-optimal in the restricted subclass of the **RGDs**.

As noted earlier, Jacroux (1985) attempted to reduce the search for optimal designs to the restricted subclass of the **RGDs**. We refer to the material presented in subsection 3.2, specifically to (3.3.14) - (3.3.17), in this context. We shall also use the same notations as in subsection 3.2. Applied to the $\mathbf{D}$-optimality criterion, this would then produce the following result:

If, for an **RGD**

$$\prod_1^{v-1} kx_{(i)} \geq \max\{m_1 m_2^{(v-3)} m_3, m_1^* \{(kA^* - m_1^*)/(v-2)\}^{(v-2)}\}$$

then the *best* **RGD** is $\mathbf{D}$-optimal in $\mathbf{D}(b,v,k)$.

As for the $\mathbf{A}$-optimality criterion, Jacroux (1985) has made a similar verification for the designs listed in John and Mitchell (1977) w.r.t. the $\mathbf{D}$-optimality criterion and arrived at the same conclusion. Again, his conclusion for the case of $b = v = 8$, $r = k = 3$ appears to be incorrect. Jacroux (1985) has also treated the case of the **SRGDs** in a similar fashion. Further, Cheng and Wu (1981) have provided a table of $\mathbf{D}$-efficiencies of some such designs.

Now we briefly describe results in Gaffke (1982) on $\mathbf{D}$-optimal designs. We omit the proofs altogether.

1. For $v = 3$ or 4 and $k = \pm 1\ (mod v)$, a design with $n_{ij} \epsilon (a, a+1)$ and $r_i \epsilon (r, r+1)$ is $\mathbf{D}$-optimal in $\mathbf{D}(b,v,k)$ where $a = [k/v]$ and $r = [bk/v]$.

2. For $v = 4$, every design which is **D**-optimal in the restricted class of binary designs is also **D**-optimal in the entire class.

3. For $v = 5$, $k = 2$, $b = 10a+s$, $a \geq 0$, a **BIBD** $(10a,5,2)$ along with the *first* s blocks of the sequence
12 34 15 23 45 13 24 25 14
is **D**-optimal in $\mathbf{D}(b,v,k)$ except when $s = 6$ in which case the six additional blocks are 12 34 15 23 14 35.

4. (a) For $v \leq 5$ and $bk = vr$, r an integer, an **RGD** always exists. Further, every **RGD** in $\mathbf{D}(b,v,k)$ is **D**-optimal in $\mathbf{D}(b,v,k)$.

   (b) For $v = 6$ and $bk = vr$, r an integer, an **RGD** always exists. Further, **D**-optimal **RGD**'s within the equireplicate class are available in all cases except $v = 6$, $b = 8$, $k = 3$. In this last case, an **RGD** is given by John and Mitchell (1977) but it is not known if it is **D**-optimal even within the equireplicate class.

6. Regular Graph Designs and John-Mitchell Conjecture

While the search for specific optimal designs continued, some common features of available optimal designs were noticed. In this process, John and Mitchell (1977) came up with a conjecture regarding the nature of optimal designs. We discuss this and other allied conjectures in this section.

We recall the definition of an **RGD** for ready reference. For given b,v and k, this is a block design for which

(i) $n_{ij} = [k/v]$ or $[k/v] + 1$, $1 \leq i \leq v$, $1 \leq j \leq b$;

(ii) r_i's are all equal;

(iii) $|\lambda_{ii'} - \lambda_{uu'}| \leq 1$ for all $i \neq i'$, $u \neq u'$.

It is evident that (ii) requires that bk/v is an integer. When this is not the case (ii) may be replaced by

(ii)$'$ $|r_i - r_{i'}| \leq 1$ for all $1 \leq i \neq i' \leq v$.

Designs satisfying (i), (ii)$'$ and (iii) have been termed *Semi Regular Graph Designs* (**SRGDs**) by Jacroux (1985). Clearly, an **SRGD** reduces to an **RGD** when bk/v is an integer. These have been termed *nearly balanced* incomplete block designs by Cheng and Wu (1981).

It may not be out of context to mention that there is a concept of *Generalized Line Graphs* (**GLGs**) introduced in Hoffman (1970). Cheng and Constantine (1986) extended it to what they termed *Regular* **GLG** designs and demonstrated that the regular **GLG** designs form an extremely useful subclass of the **RGDs** at least in respect of the **E**-optimality criterion. It is likely that such designs will also prove to be useful w.r.t. other optimality criteria.

We now address the conjecture of John and Mitchell (1977). It is based on the observation that the **RGDs** are **(M,S)**-optimal. (Vide subsection 3.3 for the definition of **(M,S)**-optimality). To see this, we note that

$$tr(k\mathbf{C}) = \sum(kr_i-\lambda_{ii}), tr(k^2\mathbf{C}^2) = \sum_i(kr_i-\lambda_{ii})^2+\sum_{i\neq i'}\sum\lambda_{ii'}^2$$

so that an **RGD** maximizes $tr(k\mathbf{C})$ and subject to this, minimizes each of $\sum_i(kr_i-\lambda_{ii})^2$ and $\sum_{i\neq i'}\sum\lambda_{ii'}^2$.

In view of Conniffe and Stone's result discussed in section 3.2 of Chapter Two, John and Mitchell (1977) put forward the following conjecture regarding the nature of an optimal design.

"If an incomplete block design is **D**-optimal (or **A**-optimal or **E**-optimal), it is an **RGD** (if an **RGD** exists)."

It was subsequently noted that in some cases a design other than an **RGD** has the same value for the smallest eigenvalue of the **C**-matrix as the best **RGD** and hence an **RGD** is not *uniquely* **E**-optimal.

It was also found that when $b = v(=n$ say) and when $k = 2$, one can construct designs which are better than *any* **RGD** w.r.t. **A**- and **E**-optimality. These are given by

$$d_{1n}: \begin{matrix} 1 & 2 & 3 & 1 & 1 & \cdots & 1 \\ 2 & 3 & 1 & 4 & 5 & \cdots & n \end{matrix}$$

$$d_{2n}: \begin{matrix} 1 & 2 & 3 & 4 & 1 & \cdots & 1 \\ 2 & 3 & 4 & 1 & 5 & \cdots & n \end{matrix}$$

For $n \geq 10$ each of the above designs is better than any **RGD** in the class $\mathbf{D}(n,n,2)$ w.r.t. **A**- and **E**-optimality and hence in these cases, an **A**- and **E**-optimal design is *not* an **RGD**. The above examples are due to Jones and Eccleston (1980) and Eccleston and Shah (1985). It may also be noted that d_{1n} and d_{2n} are *not* **(M,S)**-optimal but are superior to **(M,S)**-optimal designs w.r.t. **A**- and **E**-optimality criteria. Further, d_{1n} and d_{2n} are not equireplicate but are better than any equireplicate design in $\mathbf{D}(n,n,2)$ (which is necessarily **(M,S)**-optimal), again w.r.t. **A**- and **E**-optimality criteria.

In addition to the above, Constantine (1986) gave an example of a non-**RGD** which is again better than any **RGD** for the parameter values $b = 1350$, $v = 54$ and $k = 2$ w.r.t. **E**-optimality criterion. In fact, he gave a general method of construction of designs with $k = 2$ which are *not* **RGD**s but have better performance (w.r.t. **E**-optimality criterion) than the best available **RGD**.

Since all the above exceptions relate to designs with $k = 2$, it appears that John-Mitchell conjecture holds for designs with $k \geq 3$.

We have noted in the previous section that the **LBD**s are **A**-, **D**-, **E**-, and Φ_p optimal at least in the class of equireplicate designs. Now, an **LBD** may not be an **RGD** but this does not falsify John and Mitchell's conjecture. To see this, we note that in view of the correspondence of the eigenvalues of the **C**-matrix of a design and that of its dual, the dual of an (**M,S**)-optimal design is also (**M,S**)-optimal in the equireplicate class. Now, if an **LBD** is not an **RGD** and if one can find an **RGD** with the same values of b, v and k, the (**M,S**)-optimality of the **LBD** is violated and hence such an **RGD** does not exist. This also shows that if the dual design of an **RGD** is not an **RGD**, an **RGD** (in the corresponding design class) does not exist.

Apart from the John and Mitchell's conjecture, the following conjectures appear to be still open.

(1) Binary (or generalized binary) designs form an essentially complete class.

(2) When $k \geq 3$ and bk/v is an integer, an optimal design is necessarily equireplicate.

(3) When $k \geq 3$, an optimal design is necessarily (**M,S**)-optimal.

(4) The **LBD**s are optimal in the entire class of relevant designs.

For further conjectures, we refer to John and Williams (1982), Paterson (1983) and Wild (1987).

7. Optimal Designs With Unequal Block Sizes

Of late, there have been some studies on characterization and construction of optimal block designs with unequal block sizes. Specifically, it is assumed that the blocks in a design consist of various sizes, say $k_1, k_2, \ldots, k_s$ with b_i blocks of size k_i so that $\sum_1^s b_i = b$. Some results on universal and specific optimality of block designs with unequal block sizes have been derived in such a set-up by Pal and Pal (1988) and Lee and Jacroux (1987a, 1987b, 1987c). We remark that the optimality results are very much dependent on the specification of values of $(k_1, k_2, \ldots, k_s, b_1, b_2, \ldots, b_s)$ and as such are only moderately exciting. We omit discussion of the results. The interested reader is referred to the above cited papers.

SUMMARY OF OPTIMALITY RESULTS FOR STANDARD BLOCK DESIGNS

Designs	Optimality Criteria
BBDs	(Extended) Universal optimality
GDDs with $m = 2$, $\lambda_2 = \lambda_1+1$	Generalized optimality, **MV**-optimality
LBDs	Generalized optimality*
GDDs with $m>2$, $\lambda_2 = \lambda_1+1$	**E**- and **MV**-optimality; **A**-optimality**
GDDs with $n = 2$, $\lambda_1 = \lambda_2+1$†	**D**-, **E**- and **MV**-optimality; **A**-optimality**
GDDs with $n = 2$, $\lambda_2 = \lambda_1+1$	**E**- and **D**-optimality
GDDs with $m = 2$, $\lambda_2 = \lambda_1+2$	**E**- and **MV**-optimality$^{\psi}$
LBDs	**E**-optimality
Duals of	
GDDs with $\lambda_2 = \lambda_1+1$	**E**-optimality
GDDs with $\lambda_1 = \lambda_2+1$, $n = 2$†	**E**-optimality

*Equireplicate class
****RGD** class
$^{\psi}$provided a **GDD** with $n = 2$, $\lambda_1 = \lambda_2+1$ does *not* exist.
†$\lambda_2>0$

REFERENCES

Adhikary, B. (1965). On the properties and construction of balanced block designs with variable replications. *Cal. Statist. Assoc. Bull., 14*, 36-64.

Bagchi, S. (1988). A class of non-binary unequally replicated E-optimal designs. *Metrika, 35*, 1-12.

Cheng, C.S. (1979). Optimal incomplete block designs with four varieties. *Sankhya, Series B, 41*, 1-14.

Cheng, C.S. (1980). On the E-optimality of some block design. *J.R. Statist. Soc. B, 42*, 199-204.

Cheng, C.S. (1981). Graph and optimum design theories - some connections and examples. *Bull. Intern. Statist. Inst.* Proceedings 43rd Session (Buenos Aires) Vol. XLIX, Book 1, 580-590.

Cheng, C.S. and Constantine, G.M. (1986). On the efficiency of regular generalized line graph designs. *J. Statist. Planning and Inference, 15*, 1-10.

Cheng, C.S. and Wu, C.F. (1981). Nearly balanced incomplete block designs. *Biometrika, 68*, 493-500.

Conniffe, D. and Stone, J. (1974). The efficiency factor of a class of incomplete block designs. *Biometrika, 61*, 633-636.

Constantine, G.M. (1981). Some E-optimal block designs. *Ann. Statist., 9*, 886-892.

Constantine, G.M. (1982). On the E-optimality of PBIB designs with a small number of blocks. *Ann. Statist., 10*, 1027-1031.

Constantine, G.M. (1986). On the optimality of block designs. *Ann. Inst. Statist. Math., 38*, 161-174.

Dean, A.M. and Lewis, S.M. (1984). A comparison of upper bounds for efficiency factors of block designs. *J.R. Statist. Soc. B, 46*, 274-283.

Eccleston, J.A. and Shah, K.R. (1985). On John and Mitchell's Conjecture. Unpublished manuscript.

Fitzpatrick, S. and Jarrett, R.G. (1983). Upper bounds for the harmonic mean. Submitted.

Gaffke, N. (1982). D-optimal block designs with at most six varieties. *J. Statist. Planning and Inference, 6*, 183-200.

Hedayat, A. and Federer, W.T. (1974). Pairwise and variance balanced incomplete block designs. *Ann. Inst. Statist. Math., 26*, 331-338.

Hoffman, A.J. (1970). $-1-\sqrt{2}$? In: R. Guy Ed. *Combinatorial Structures and Their Applications*. Gordon and Breach, New York, 173-176.

Jacroux, M. (1980a). On the E-optimality of regular graph designs. *J.R. Statist. Soc. B, 42*, 205-209.

Jacroux, M. (1980b). On the determination and construction of E-optimal block designs with unequal number of replicates. *Biometrika, 67*, 661-667.

Jacroux, M. (1982). Some E-optimal designs for the one-way and two-way elimination of heterogeneity. *J.R. Statist. Soc. B, 44*, 253-261.

Jacroux, M. (1983a). On the E-optimality of block designs. *Sankhya, Series B, 45*, 351-361.

Jacroux, M. (1983b). Some minimum variance block designs for estimating treatment differences. *J.R. Statist. Soc. B, 45*, 70-76.

Jacroux, M. (1984). On the D-optimality of group divisible designs. *J. Statist. Planning and Inference, 9*, 119-129.

Jacroux, M. (1985). Some sufficient conditions for the type I optimality of block designs. *J. Statist. Planning and Inference, 11*, 385-394.

James, A.T. and Wilkinson, G.N. (1971). Factorization of the residual operator and canonical decomposition of non-orthogonal factors in the analysis of variance. *Biometrika, 58*, 279-294.

Jarrett, R.G. (1977). Bounds for the efficiency factor of block designs. *Biometrika, 64*, 67-72.

Jarrett, R.G. (1983). Definitions and properties for m-concurrence designs. *J.R. Statist. Soc. B, 45*, 1-10.

John, J.A. and Mitchell, T.J. (1977). Optimal incomplete block designs. *J.R. Statist. Soc. B, 39*, 39-43.

John, J.A. and Williams, E.R. (1982). Conjectures for incomplete block designs. *J. R. Statist. Soc. B, 44*, 221-225.

Jones, B. and Eccleston, J.A. (1980). Exchange and interchange procedures to search for optimal designs. *J. R. Statist. Soc. B, 42*, 238-243.

Kageyama, S. (1980). On properties of efficiency balanced designs. *Comm. Statist.*

Theor. Meth. A, 9, 597-616.

Kempthorne, O. (1956). The efficiency factor of an incomplete block design. *Ann. Math. Statist., 27*, 846-849.

Lee, K.Y. and Jacroux, M. (1987a). Some sufficient conditions for the *E*- and *MV*-optimality of block designs having unequal block sizes. *Ann. Inst. Statist. Math., 39*, 385-397.

Lee, K.Y. and Jacroux, M. (1987b). On the construction of *E*- and *MV*-optimal group divisible designs with unequal block sizes. *J. Statist. Planning and Inference, 16*, 193-201.

Lee, K.Y. and Jacroux, M. (1987c). On the *E*-optimality of block designs having unequal block sizes. *Sankhya (B), 49*, 126-136.

Pal, S. and Pal, S. (1988). Non-proper variance balanced designs and optimality. *Comm. Statist. Theo. and Methods, 17*, 1685-1695.

Paterson, L.J. (1983). Circuits and efficiency in incomplete block designs. *Biometrika, 70*, 215-225.

Pearce, S.C. (1968). The mean efficiency of equireplicate designs. *Biometrika, 55*, 251-253.

Pearce, S.C., Calinski, T. and Marshall, T.F. de C. (1974). The basic contrasts of an experimental design with special reference to the analysis of data. *Biometrika, 61*, 449-460.

Puri, P.D. and Nigam, A.K. (1975). On patterns of efficiency-balanced designs. *J.R. Statist. Soc. B, 37*, 457-458.

Puri, P.D. and Nigam, A.K. (1977). Balanced block designs. *Comm. Statist. - Theor. Meth. A, 6*, 1171-1179.

Rao, C.R. (1947). General methods of analysis for incomplete block designs. *J. Amer. Statist. Assoc., 42*, 541-561.

Sathe, Y.S. and Bapat, R.B. (1985). On the *E*-optimality of truncated BIBD. *Cal. Statist. Assoc. Bull., 34*, 113-117.

Sinha, B.K. and Shah, K.R. (1988). Optimality aspects of 3-concurrence most balanced designs. *J. Statist. Planning and Inference, 20*, 229-236.

Takeuchi, K. (1961). On the optimality of certain type of PBIB designs. *Rep. Statist. Appl. Res. Un. Japan Sci. Engrs., 8* (No. 3) 140-145.

Takeuchi, K. (1963). A remark added to "On the optimality of certain type of PBIB designs". *Rep. Statist. Appl. Res. Un. Japan Sci. Engrs., 10* (No. 3) 47.

Tocher, K.D. (1952). The design and analysis of block experiments (with discussion). *J.R. Statist. Soc. B, 14*, 45-100.

Wild, P. (1987). On circuits and optimality conjectures for block designs. *J. R. Statist. Soc., B, 49*, 90-94.

Williams, E.R. and Patterson, H.D. (1977). Upper bounds for efficiency factors in block designs. *Austral. J. Statist., 19*, 194-201.

CHAPTER FOUR

ROW-COLUMN DESIGNS

1. Introduction

In this Chapter, we will discuss mostly the optimality aspects of row-column designs. The related constructional aspects will *not* be dealt with rigorously but some references will be given.

Our set-up is as follows. There are bk experimental units arranged in the form of a $k \times b$ array. A design involving v treatments consists of an arrangement of the treatments to the bk experimental units. It is assumed that heterogeneity occurs among these experimental units across the rows as well as across the columns. The designs are meant to eliminate heterogeneity in both the directions while providing information on the treatment effects. The usual (fixed effects) linear model specifies the expectation of an observation in the (i,j)th cell and under the hth treatment as $\mu+\beta_i+\gamma_j+\tau_h$ where μ, β_i, γ_j and τ_h stand respectively for the general effect, ith row effect, jth column effect and hth treatment effect. Further, it is assumed that all the bk observations are uncorrelated with the common unknown variance σ^2. In such a situation two types of incidence matrices emerge: *treatments* vs. *rows* and *treatments* vs. *columns*. These are denoted by $\mathbf{M}(v \times k)$ and $\mathbf{N}(v \times b)$ respectively. As before, we are interested in linear inference problems involving the treatment effects whose information matrix is shown in (1.2.5) of Chapter One. For ready reference, we rewrite it below.

$$\mathbf{C} = r^{\delta} - k^{-1}\mathbf{NN}' - b^{-1}\mathbf{MM}' + b^{-1}k^{-1}\mathbf{rr}' \tag{4.1.1}$$

where $\mathbf{r} = \mathbf{N1} = \mathbf{M1}$, $\mathbf{r}^{\delta} = Diag(r_1, r_2, \ldots, r_v)$. Further, $\mathbf{r}'\mathbf{1} = bk$. Explicitly written,

$$C_{hh'} = r_h \delta_{hh'} - \frac{\lambda_{hh'}}{k} - \frac{\mu_{hh'}}{b} + \frac{r_h r_{h'}}{bk} \tag{4.1.2}$$

where

$$\left.\begin{array}{l} \delta_{hh'} = 1 \text{ if } h = h'; = 0 \text{ otherwise} \\ ((\lambda_{hh'})) = \mathbf{NN}' \\ ((\mu_{hh'})) = \mathbf{MM}' \end{array}\right\} \tag{4.1.3}$$

The $\mathbf{C}$-matrix has all the usual properties of being symmetric nnd with row and column sums zeros. For *all* treatment contrasts to be estimable, it is n.s. that the underlying design be connected i.e., rank $(\mathbf{C}) = v-1$. In the sequel, we will work only with connected designs. However, as remarked earlier (in section 2, Chapter One), connectedness is *not* always easy to verify in a row-column design.

For given b, v and k, a design is called a *Generalized Youden Design* (**GYD**) if the following hold:

(i) $m_{hi} = \frac{b}{v}$ in case b/v is an integer

$= [b/v]$ or $[b/v]+1$, otherwise; $1 \le h \le v$, $1 \le i \le k$

(ii) $n_{hj} = k/v$ in case k/v is an integer

$= [k/v]$ or $[k/v]+1$, otherwise; $1 \le h \le v$, $1 \le j \le b$

(iii) r_h's are all equal, $\lambda_{hh'}$'s are all equal and $m_{hh'}$'s are all equal $(1 \le h \ne h' \le v)$.

Clearly, for a **GYD**, $r_1 = r_2 = \cdots = r_v = bk/v = r$ (say).

Basically, a **GYD** is a design which is *balanced* in both directions i.e., it is a **BBD** when considered as a block design with the rows or columns as blocks. Thus for a **GYD**, λ_{hh}'s are all equal and μ_{hh}'s are all equal. That is, the **C**-matrix has all diagonal elements equal and also all off-diagonal elements equal. In other words, it is completely symmetric (c.s.). When $b = v = k$, a **GYD** reduces to what is known as a *Latin Square Design* (**LSD**) and for $b = v > k$, a **GYD** is simply a *Youden Square Design* (**YSD**). A **GYD** is called *regular* when at least one of b/v or k/v is an integer. Otherwise, it is called *non-regular*. Observe that **LSDs** and **YSDs** are regular. We refer to Raghavarao (1971) for an account of the **LSDs** and the **YSDs**. For various combinatorial and constructional aspects of the **GYDs**, we refer to Kiefer (1975), Ruiz and Seiden (1974) and Ash (1981). Another related paper is due to Sinha (1980). When $b = k$, a design with $(\mathbf{M}\vdots\mathbf{N})$ forming the incidence matrix of a **BBD** with $2b$ blocks each of size b has been termed a *Pseudo Youden Design* (**PYD**) by Cheng (1981) who has discussed the optimality and constructional aspects of such designs.

Kiefer (1975) succeeded in establishing universal optimality of the regular **GYDs** and specific optimality of the non-regular **GYDs** w.r.t. the **A**-, **D**- and **E**-optimality criteria (except for $v = 4$ for **D**-optimality) by considering the entire class $\mathbf{D}(b,v,k)$ of connected designs as forming the competing designs. Below we will mainly follow his approach and present the optimality results for the **GYDs** and the **PYDs** in sections 2 and 3.

Most of the available optimality results relate to the **GYDs** or **PYDs**. In section 4 we give bounds for the efficiency factor for any row-column design and present two theorems which establish the optimality of certain designs having orthogonality for rows and columns *adjusted* for treatments. We conclude this section with some examples of designs which are either optimal or are expected to be so.

Recently, Saharay (1986) studied some optimality aspects of row-column designs with observations along a transversal missing. Universal and specific optimality of some classes of designs have been established in this set-up.

2. Universal Optimality of the Regular GYDs

In the set-up of a row-column design, for given b, v and k, suppose a regular **GYD** exists. We want to establish its universal optimality in the extended sense as described in subsection 3.2 of Chapter One. Since the **C**-matrix of a **GYD** is completely symmetric, we may refer directly to Proposition 1 (restated) in subsection 3.2 of Chapter One and observe that it is enough to verify that the **C**-matrix of a regular **GYD** possesses maximum trace among all competing designs in $\mathbf{D}(b,v,k)$. This is exactly what is shown below. From (4.1.2),

$$\sum_h C_{hh} = bk - \frac{1}{b}\sum_i\sum_h m_{hi}^2 - \frac{1}{k}\sum_j\sum_h n_{hj}^2 + (\sum_h r_h^2)/bk$$

so that

$$(bk-\sum_h C_{hh})bk = k\sum_i\sum_h m_{hi}^2 + b\sum_j\sum_h n_{hj}^2 - \sum_h r_h^2. \tag{4.2.1}$$

We want to show that the RHS of (4.2.1) is minimized for the regular **GYD**. Observe that $\sum_h m_{hi} = b$, $1 \le i \le k$ and $\sum_h n_{hj} = k$, $1 \le j \le b$. Since it is a regular set-up, we assume that k/v is an integer. Now

$$b\sum_j\sum_h n_{hj}^2 - \sum_h r_h^2 = b\sum_h\{\sum_j n_{hj}^2 - \frac{r_h^2}{b}\} = b\sum_h[\sum_j\{n_{hj} - \sum_j n_{hj}/b\}^2] \ge 0$$

and the equality occurs for a design with $n_{hj} = k/v$, $r_h = bk/v$, $1 \le j \le b$, $1 \le h \le v$. Clearly, this is the case with the **GYD**. The rest is to show that $\sum_i\sum_h m_{hi}^2$ also attains its minimum for the **GYD**. Since $\sum_h m_{hi} = b$, $\sum_h m_{hi}^2$ attains its lower bound b^2/v when $v|b$ and $m_{hi} = b/v$, $1 \le h \le v$. In case $v \not| \, b$, the minimum is reached when $m_{hi} = [b/v]$ or $[b/v]+1$, $1 \le i \le k$, $1 \le h \le v$. For fixed i, let f_i be the number of m_{hi}'s that are equal to $[b/v]$ so that the rest are equal to $[b/v]+1$. Thus, $b = f_i[b/v] + (v-f_i)([b/v]+1)$ i.e., $f_i = v[b/v] - b + v$ for all i. Hence $\sum_i\sum_h m_{hi}^2 \ge k[\{v([b/v]+1)-b\}[b/v]^2 + (b-v[b/v])([b/v]+1)^2]$. It is now easy to check that this equality is attained by many designs and, in particular, by the **GYD**. Since for the **GYD**, $bk(bk-\sum C_{hh})$ attains lower bounds to both the parts of the RHS expression in (4.2.1), it maximizes $tr(\mathbf{C})$ among all designs in this set-up. This establishes (extended) universal optimality of the regular **GYD**s.

Remark 4.2.1. If neither b/v nor k/v is an integer, the above argument breaks down since a **GYD** does not necessarily maximize $tr(\mathbf{C})$**. Proposition 1 (restated) of Chapter One thus fails to be applicable generally in such a nonregular set-up and so we cannot claim universal optimality of the **GYD**s. This does not, however, mean that the **GYD**s are *not* optimal w.r.t. some specific optimality criteria. As a matter of fact, they are still **A**-, **D**- and **E**-optimal (in the sense of (1.3.3), (1.3.4) and (1.3.6)) whatever b, v and k (with the exception of $v=4$ as regards the **D**-optimality criterion). Naturally, new tools are needed for establishing these specific optimality results which we present below in section 3. We mention in passing that as the **GYD**s possess completely symmetric **C**-matrices, **D**-optimality (**A**-optimality) of such a design would automatically imply its **A**- and **E**-optimality

** An example to this effect may be found in Kiefer (1958). See also Remark 4.3.2.

(respectively, E-optimality). This follows from the general observation made in subsection 3.3 of Chapter One (immediately after (1.3.6)). However, E-optimality results are easier to derive and they cover more combinations of (b,v,k) values. Moreover, the technique employed may be of independent interest.

3. Nonregular GYDs: Specific Optimality Results

3.1 E-Optimality of Non-Regular GYDs

We are in a non-regular set-up so that $v \nmid k$ and $v \nmid b$. However, we assume the existence of a **GYD** so that $v \mid bk$. Clearly, the C-matrix underlying the **GYD** is completely symmetric but it does *not* necessarily maximize the trace. We intend to show that the **GYDs** are still E-optimal. The approach is to obtain an upper bound to the smallest non-zero eigenvalue $kx_{(1)}$ for an arbitrary design in this set-up and to show that this bound is attained by the smallest non-zero eigenvalue of the C-matrix underlying the **GYD**. Towards this, we use the bound

$$x_{(1)} \leq \frac{v}{v-1} \min_h C_{hh} \tag{4.3.1}$$

obtained by using Constantine's *averaging technique* elaborated in subsection 2.2 of Chapter Three. Note that equality is reached in (4.3.1) for a **GYD**. Next, we obtain an upper bound to C_{hh} for an arbitrary design in $\mathbf{D}(b,v,k)$ for fixed value of r_h for a given h. Referring to (4.1.2), one can show that

$$C_{hh} \leq (r_h + \frac{r_h^2}{bk}) - \frac{1}{b}\{r_h(2[\frac{r_h}{k}] + 1) - k[\frac{r_h}{k}] - k[\frac{r_h}{k}]^2\}$$

$$- \frac{1}{k}\{r_h(2[\frac{r_h}{b}] + 1) - b[\frac{r_h}{b}] - b[\frac{r_h}{b}]^2\} = \phi(r_h) \text{ (say)} \tag{4.3.2}$$

It is clear that for the **GYD**, $C_{hh} = \phi(r)$ for all h. Further, for any competing design $r_h \leq r = bk/v$ for *some* h. It is then enough to show that $\phi(r_h) \leq \phi(r)$ for $r_h \leq r$. The case of equality is trivial and so we assume $r_h < r$. As a matter of fact, we will establish $\phi(x)\uparrow$ in x for all integral $x<r$. For this, first observe that $\min\{\sum_i m_{hi}^2\}$ subject to $\sum_i m_{hi} = x+1$ is equal to $\min\{\sum_i m_{hi}^2\}$ subject to $\sum_i m_{hi} = x$ *plus* $\{[x/k]+1\}^2 - [x/k]^2 (= 1+2[x/k])$. Using this, we readily compute

$$\phi(x+1) - \phi(x) = 1 + \frac{1+2x}{bk} - \frac{1}{b}(1+2[x/k]) - \frac{1}{k}(1+2[x/b]) \tag{4.3.3}$$

and for this to be positive, a sufficient condition is that $1 + \frac{1}{bk} - \frac{1}{b} - \frac{1}{k} \geq \frac{2x}{bk}$ i.e., $(b-1)(k-1) \geq 2x$ which is true for $x<r = \frac{bk}{v}$ whenever $b,k>v\geq 4$. This is exactly the situation in this nonregular case (the cases of $v = 2,3$ are regular; in fact, the cases of $v =$ prime, are all regular; in any nonregular case, we necessarily have $b,k>v$

since the **GYD** is assumed to exist). Thus, within the class of all connected designs for given b,v and k, a **GYD**, whenever it exists, is **E**-optimal.

Remark 4.3.1. It is easy to see that in case $b = k$, $C_{hh} = \phi(r)$ for all h for a **PYD** which is, then, **E**-optimal. This is mentioned in Cheng (1981).

3.2 A- and D-Optimality of Nonregular GYDs

The context is the same as in subsection 3.1. Remark 4.2.1 points out inapplicability of Proposition 1 (restated) in this nonregular set-up. We need new techniques to establish **A**- and **D**-optimality of the **GYDs**. We proceed as follows.

The arguments to be developed will be the same for a general class of optimality criteria (including these two) $\psi_f(x_1,x_2,\ldots,x_{v-1}) = \sum_1^{v-1} f(x_i)$ where f is convex and assumed to be nonincreasing over $[0,\infty)$. The problem is to check whether a **GYD** becomes a ψ_f-optimal design (which seeks to minimize ψ_f). Towards this, we first observe the following.

Lemma 4.3.1. For every connected design in $\mathbf{D}(b,v,k)$,

$$\sum_1^{v-1} f(x_i) \geq \frac{v-1}{v}\sum_1^{v} f\left(\frac{v}{v-1}C_{hh}\right) \tag{4.3.4}$$

with "=" iff x_i's are all equal.

Proof. Suppose $\mathbf{Q}(v\times v)$ diagonalizes $\mathbf{C}$ so that we write $\mathbf{C} = \mathbf{Q'DQ}$ with $\mathbf{Q} = ((q_{hh'}))$ where $q_{v1} = q_{v2} = \cdots = q_{vv} = \frac{1}{\sqrt{v}}$ and with $\mathbf{D} = diag.(x_1,x_2,\ldots,x_{v-1},0)$. This yields $C_{hh} = \sum_{i=1}^{v-1} q_{hi}^2 x_i$, $1\leq h\leq v$. Further, $\sum_h q_{hi}^2 = 1$, $1\leq i\leq v-1$ and $\sum_{i=1}^{v-1} q_{hi}^2 = \frac{v-1}{v}$, $1\leq h\leq v$. Hence, f being convex,

$$\left(\frac{v-1}{v}\right)\sum_{h=1}^{v} f\left(\frac{v}{v-1}C_{hh}\right) = \left(\frac{v-1}{v}\right)\sum_{h=1}^{v} f\left(\frac{v}{v-1}\sum_i q_{hi}^2 x_i\right)$$

$$\leq \left(\frac{v-1}{v}\right)\sum_h \frac{v}{v-1}\sum_i q_{hi}^2 f(x_i) = \sum_i f(x_i)\left(\sum_h q_{hi}^2\right) = \sum f(x_i).$$

The case of equality is easy to check.

In view of (4.3.4), it is enough to check that $\sum_h f\left(\frac{v}{v-1}C_{hh}\right)$ attains its minimum for the **GYD**. Again (4.3.2) implies $C_{hh} \leq \phi(r_h)$ for all h and since f is nonincreasing, one gets

$$\sum_1^{v-1} f(x_i) \geq \left(\frac{v-1}{v}\right)\sum_1^{v} f\left(\frac{v}{v-1}\phi(r_h)\right) \tag{4.3.5}$$

where "=" holds iff all the x_i's are equal. The problem thus reduces to checking that $\sum_h f\left(\frac{v}{v-1}\phi(r_h)\right)$ attains its minimum for the **GYD**. In effect, we have to examine the validity of

$$\sum f(\frac{v}{v-1}\phi(r_h)) \geq vf(\frac{v}{v-1}\phi(r)). \tag{4.3.6}$$

Looking at the expression for ϕ in (4.3.2), one feels that it is easy to handle

$$g(x) = bk\phi(x) \tag{4.3.7}$$

Further, we define

$$q(x) = -f(\frac{v}{v-1}\phi(x)) \tag{4.3.8}$$

Once for all we make it clear that x will represent only non-negative integral values. Clearly, (4.3.6) would follow at once if $q(x)$ were concave. Unfortunately, however, this is not always the case. This is what motivated Kiefer (1975) for further development of new tools.

Let $\mathbf{G} = \{0,1,2, \ldots, bk\}$ and $\bar{q}(x)$ be the concave envelope of $q(x)$ i.e., the smallest function $\geq q(x)$ whose second differences $\bar{q}(x+2) + \bar{q}(x) - 2\bar{q}(x+1)$ are all ≤ 0. Then (4.3.6) will still hold if

$$\bar{q}(r) = q(r) \tag{4.3.9}$$

We show that a sufficient condition for (4.3.9) to hold is that

$$q(x+1) - q(x) \uparrow \text{in } x \text{ for } c_0 \leq x < d_0 \tag{4.3.10}$$

where $[c_0,d_0]$ represents a very special subset of $\mathbf{G}$ to be described below. To establish this result, we need first a detailed analysis of the nature of the function $g(x)$ defined in (4.3.7).

Let us write $[c,d]$ for an interval of successive integers. Let $\mathbf{U} = \{n: 0 \leq n \leq bk,\ n$ is an integral multiple of b or $k\}$ and let $\mathbf{V} = \{n: n \epsilon \mathbf{U},\ n \leq \frac{bk}{2}\}$. If $c,d \epsilon \mathbf{U}$, $c<d$ and no integer between c and d is in $\mathbf{U}$, we call $[c,d]$ an elementary interval. The elementary interval $[c_0,d_0]$ containing the integer $r(= \frac{bk}{v})$ is called the *basic* interval. Since the set-up is nonregular, we note that $c_0<r<d_0$ and, hence, $1 + c_0 \notin \mathbf{U}$, $r \notin \mathbf{U}$ and, further, it can be shown that $d_0 \notin \mathbf{V}$. The function $g(x)$ is observed to share the following properties (we omit the proofs which may be supplied without much difficulty):

(a) Within each elementary interval $[c,d]$, $\Delta(x) = g(x+1) - g(x)$ is linear in x and increasing for $c \leq x < d$ i.e., g is a convex quadratic on each elementary interval.

(b) g is increasing in each elementary interval $[c,d]$ with $d \leq d_0$ (this follows from (a) and the fact that $\phi(x)$ is increasing in x for $x<r$, a result established in connection with the proof of **E**-optimality in the previous subsection).

(c) g is symmetric about $\frac{bk}{2}$.

(d) If $c_1, c_2 \epsilon \mathbf{U}$ with $c_1<c_2$, then $\Delta(c_1) \geq \Delta(c_2)$ and, consequently, $\Delta(c_1-1) \geq \Delta(c_2-1)$.

(e) g is nondecreasing on $\mathbf{V}$ and nonincreasing on the remainder of $\mathbf{U}$.

Suppose now that $\bar{d} \epsilon \mathbf{G}$ is the first integer where g attains its maximum. Clearly, by (e), $\bar{d} \epsilon \mathbf{V}$ and, by (b), $\bar{d} \geq d_0$. Since $-f$ in (4.3.8) is monotone, to prove (4.3.9) it is enough to consider

$$\mathbf{G}' = \{x: x \epsilon \mathbf{G}, x \leq \bar{d}\}.$$

By (e) above, in any elementary interval $[c,d]$ of $\mathbf{G}'$, $g(c) \leq g(d)$ and, consequently, $q(c) \leq q(d)$. Again, in any such interval $[c,d]$ where g is not monotone so that there exists $x' \epsilon [c,d]$ for which $g(x') \leq g(c)$, we have $q(x') \leq q(c) \leq q(d)$. Consequently, in verifying (4.3.9), $\mathbf{G}'$ can be replaced by its subset $\mathbf{G}''$ obtained by excluding such points x' so that now for any two points x_1 and x_2 of $\mathbf{G}''$, we have $q(x_1) \leq q(x_2)$ whenever $x_1 \leq x_2$. Next one can see that a sufficient condition for (4.3.9) to hold using $\mathbf{G}''$ is that

$$q(x_1) + q(x_2) \leq q(x_1+1) + q(x_2-1)$$
$$\text{i.e., } q(x_1+1) - q(x_1) \geq q(x_2) - q(x_2-1) \tag{4.3.11}$$

whatever $x_1+1 \leq r \leq x_2-1$, x_1, $x_2 \epsilon \mathbf{G}''$.

We may note that x_2-1 need not belong to $\mathbf{G}''$. One can establish (4.3.11) by proving that

$$\left.\begin{array}{ll} \text{(i)} & \min\limits_{0 \leq x_1 < r} [q(x_1+1)-q(x_1)] = q(r)-q(r-1) \\ \text{(ii)} & \max\limits_{r<x_2 \epsilon \mathbf{G}''} [q(x_2)-q(x_2-1)] = q(r+1)-q(r) \\ \text{(iii)} & q(r+1) - q(r) \leq q(r) - q(r-1). \end{array}\right\} \tag{4.3.12}$$

Thus, in effect, (4.3.12) implies (4.3.9). We shall now show that (4.3.12) follows from (4.3.10) which refers to concavity of q within $[c_0,d_0]$ where $[c_0,d_0]$ is the basic interval. To see this, we observe readily that (4.3.10) implies (4.3.12) (iii). Now let $[c,d]$ be any elementary interval to the left of $[c_0,d_0]$ so that $d \leq c_0$. Let $x_1 \epsilon [c,d]$ so that $x_1<d \leq c_0$. By the property (d) above, one gets $\Delta(d) \geq \Delta(c_0)$ as also $\Delta(c) \geq \Delta(d)$.

Again, by the property (a), $\Delta(x_1) \geq \Delta(c)$ and, hence, $\Delta(x_1) \geq \Delta(d) \geq \Delta(c_0)>0$ (as is evident from the property (b) above) i.e., $g(x_1+1) - g(x_1) \geq g(c_0+1) - g(c_0)>0$. Let $y_1 = g(x_1+1)$, $y_2 = g(x_1)$, $y_3 = g(c_0+1)$ and $y_4 = g(c_0)$. Then, by the property (b) above, $y_2<y_1 \leq y_4<y_3$ and we have further deduced above that $y_1-y_2 \geq y_3-y_4$. Now since $-f$ is

concave nondecreasing, one must have

$$\frac{f(y_2')-f(y_1')}{y_1'-y_2'} \geq \frac{f(y_4')-f(y_3')}{y_3'-y_4'} \quad \text{where } y_i' = \frac{v}{v-1} \cdot \frac{y_i}{bk}$$

i.e., $q(x_1+1) - q(x_1) \geq (\frac{y_1-y_2}{y_3-y_4})(q(c_0+1) - q(c_0))$ which implies $q(x_1+1) - q(x_1) \geq q(c_0+1) - q(c_0)$. It is now clear that this, combined with (4.3.10), yields (4.3.12) (i). By an analogous argument, one can establish (4.3.12) (ii) starting with (4.3.10). Of course, (4.3.12) (iii) is included in the description of (4.3.10).

Thus, in order to establish ψ_f-optimality of the **GYDs**, it is enough to establish (4.3.10) with $q(x) = -f(g(x))$. This is all what we do now in getting into the **A**- and **D**-optimality properties of the **GYDs** for given b, v and k in a non-regular set-up i.e., with $v \not| k$ and $v \not| b$.

A-optimality of nonregular GYDs

Here $f(x) = \frac{1}{x}$ so that $q(x) = -\frac{1}{g(x)}$. Now condition (4.3.10) can be written as

$$\frac{1}{g(x+1)} + \frac{1}{g(x-1)} - \frac{2}{g(x)} \geq 0 \ \text{ for all } c_0<x<d_0 \tag{4.3.13}$$

This is equivalent to $\Gamma(x) \geq 0$ where

$$\Gamma(x) = \frac{1}{2}g(x)[\Delta(x-1) - \Delta(x)] + \Delta(x-1)\Delta(x) \tag{4.3.14}$$

with $\Delta(x) = g(x+1) - g(x)$. Over the interval $[c_0,d_0]$, the function $g(x)$ (by (a) and (b) above) is of the form $\alpha + \beta x + x^2$ with $\beta + 2x \geq 0$ for $x \geq c_0+1$. This gives

$$\Gamma(x) = 3x^2 + 3\beta x + (\beta^2-\alpha-1) \tag{4.3.15}$$

and for $x \geq c_0+1$, $\frac{d\Gamma(x)}{dx} \geq 0$. Hence, our result will follow if we can show that $\Gamma(c_0+1) \geq 0$. This part is relatively easier (Vide Kiefer (1975)) and we omit the details (more because of our intention to work out all the details of calculations needed for the **D**-optimality proof). We simply note that the result holds for all b, v and k with $v \geq 4$.

D-optimality of nonregular GYDs with v ≥ 6

Here $f(x) = -log\ x$ so that $q(x) = \log g(x)$. Hence condition (4.3.10) becomes

$$g^2(x) \geq g(x+1)g(x-1) \text{ for } c_0<x<d_0. \tag{4.3.16}$$

As before, writing $\Delta(x) = g(x+1) - g(x)$ and $\Delta(x-1) = g(x) - g(x-1)$, (4.3.16) becomes

$$g(x)\{\Delta(x-1) - \Delta(x)\} + \Delta(x)\Delta(x-1) \geq 0 \text{ for } c_0<x<d_0.$$

Let $\Gamma_0(x) = g(x)\{\Delta(x-1)-\Delta(x)\} + \Delta(x)\Delta(x-1)$. Using the fact that $g(x) = \alpha + \beta x + x^2$ over $c_0 \leq x \leq d_0$, one readily gets

$$\Gamma_0(x) = 2x^2 + 2\beta x + (\beta^2 - 2\alpha - 1) \tag{4.3.17}$$

and for $x \geq c_0+1$, $\dfrac{d\Gamma_0(x)}{dx} = 4x+2\beta \geq 0$. Hence, the result will follow if we can show that $\Gamma_0(c_0+1) \geq 0$ (which becomes possible *whenever* $v \geq 6$). It involves many steps and we are going to present all the details this time.

Step 1. Without any loss of generality, we set $k \leq b$ and $c_0 = kK \geq bB$ where $B = [c_0/b]$. Thus, for $c_0 \leq x < d_0$, $[x/k] = K$ and $[x/b] = B$. Hence, from (4.3.2), (4.3.7) and the form $\alpha + \beta x + x^2$ of $g(x)$, one derives readily

$$\left.\begin{aligned} \alpha &= k^2K(K+1) + b^2B(B+1) \\ \beta &= bk - (2K+1)k - (2B+1)b \end{aligned}\right\} \tag{4.3.18}$$

Now (4.3.18) shows that both β and $-\alpha$ are decreased if we replace B by $\frac{kK}{b}$ $(\geq B)$. The resulting decreased value of β is $bk - (2K+1)k - (2kK+b) = bk - 4c_0 - (b+k)$ which is positive since $c_0 < r = \frac{bk}{v} \leq \frac{bk}{6}$ for $v \geq 6$ (the set-up is nonregular so that $b,k > v \geq 6$). Thus β^2 is also decreased for $B = \frac{kK}{b}$. Writing $bk = \Pi$ and $b+k = \epsilon$, we thus derive, from (4.3.7) (and upon simplification)

$$\begin{aligned} \Gamma_0(c_0+1) &\geq 2(c_0+1)^2 + 2(c_0+1)(\Pi - 4c_0 - \epsilon) + \{(\Pi - 4c_0 - \epsilon)^2 - 4c_0^2 - 2c_0\epsilon - 1\} \\ &= 6c_0^2 + (4\epsilon - 4 - 6\Pi)c_0 + (1+\Pi-\epsilon)^2 = Q(c_0) \text{ (say)} \end{aligned}$$

where

$$Q(z) = 6z^2 + (4\epsilon - 4 - 6\Pi)z + (1+\Pi-\epsilon)^2. \tag{4.3.19}$$

We will show that $Q(c_0) \geq 0$.

Step 2. Since the set-up is nonregular, we may write $v = v_1v_2$, $b = v_1b'$ and $k = v_2k'$. Then $c_0 = k[b/v] = k[b'/v_2] = v_2k'[b'/v_2] \leq v_2k'(\frac{b'-1}{v_2}) = b'k' - k' = \frac{bk}{v} - k'$. Again,

$k>v = v_1v_2$ and $k = v_2k' \Rightarrow k \geq v_2(1+v_1) = v_2+v$ i.e., $k-v \geq v_2$. Hence, $c_0 \leq \frac{bk}{v} - k' = \frac{bk}{v} - \frac{k}{v_2} \leq \frac{bk}{v} - \frac{k}{k-v} \leq \frac{bk}{6} - \frac{k}{k-6}$ since $v \geq 6$[1].

We now verify that $Q'(z)<0$ for $z \leq \frac{bk}{6}$. We have $Q'(z) = 12z + (4\epsilon-4-6\Pi)$ and so $Q'(bk/6) = 2bk + 4\epsilon-4-6\Pi = -4\Pi-4+4\epsilon = -4(\Pi-\epsilon+1) = -4(b-1)(k-1)<0$. Thus the verification is done. Since $c_0 \leq \frac{bk}{6} - \frac{k}{k-6} < \frac{bk}{6}$, it is enough to verify now that $Q(\frac{bk}{6} - \frac{k}{k-6}) \geq 0$. This we do in the next step.

Step 3. We get (after a little simplification)

$$6(k-6)^2Q(\frac{bk}{6} - \frac{k}{k-6}) = (k-6)^2\Pi^2 + \Pi(k-6)\{24k-8(\epsilon-1)(k-6)\}$$
$$+ \{6k-2(k-6)(\epsilon-1)\}^2 + 2(\epsilon-1)^2(k-6)^2$$

and this RHS expression is a quadratic in Π whose derivative w.r.t. Π is $2(k-6)^2\Pi + (k-6)\{24k-8(\epsilon-1)(k-6)\}$. In this nonregular set-up with $v \geq 6$, we have $b \geq k \geq 8$ so that $\Pi \geq 8(\epsilon-8)$ and, further, $\epsilon = b+k \geq 2k \geq 16$. These considerations may be easily applied to justify positivity of the above derivative. Hence, we substitute for Π the value of $8(\epsilon-8)$ in the RHS expression above and establish its nonnegativity. The resulting expression is a polynomial in k and ϵ and is given by

$$\psi(k,\epsilon) = 64(k-6)^2(\epsilon-8)^2 + 8(k-6)(\epsilon-8)\{24k-8(\epsilon-1)(k-6) +$$
$$\{6k-2(k-6)(\epsilon-1)\}^2 + 2(k-6)^2(\epsilon-1)^2 = \{6k-2(k-6)(\epsilon-1)\}^2 +$$
$$896(k-6)(2k-21) + 2(k-6)(\epsilon-1)\{(k-6)(\epsilon-1)-64(2k-21)\} \quad \text{(after simplification)}.$$

The case $2k<21$ is disposed of right now since $\epsilon-1>7$. Otherwise, we rewrite ψ as

$$\psi(k,\epsilon) = 6(k-6)^2(\epsilon-1)^2 - 8(k-6)(\epsilon-1)\{3k+16(2k-21)\}$$
$$+ 36k^2 + 896(k-6)(2k-21).$$

As a function of $u = (k-6)(\epsilon-1)$, ψ is a quadratic in u ($\psi = au^2+bu+c$ with $a = 6$) and its zeros have opposite signs. Thus it suffices to verify that u is at least equal to the positive root. This we do in the next step.

Step 4. Since $b = 24k+128\ (2k-21)$ and $c = 36k^2 + 896(k-6)(2k-21)$ and since $\epsilon \geq 2k$, all we want to show now is that

$$(2k-1)(k-6) \geq \frac{1}{12}\{b+\sqrt{b^2-24c}\}$$

i.e, $12(2k-1)(k-6) - b \geq \sqrt{b^2-24c}$

---

[1]See Sinha (1980) in this context for combinatorial inequalities.

i.e., $(24k^2-436k+2760)^2 + 24\{36k^2+896(k-6)(2k-21)\} \geq (280k-2688)^2$

which will be a consequence of $24k^2-436k+2760 \geq (280k-2688)$ (both sides are positive since now $2k>21$) i.e., of $24k^2-716k+5448 \geq 0$ whatever $k \geq 8$. This is, of course, true and we may stop now.

Thus we have demonstrated **D**-optimality of the **GYD**s for all $v \geq 6$.

Remark 4.3.2. All the results so far discussed in this Chapter are due to Kiefer (1958, 1959, 1975); see also Kiefer (1971). In an early paper, Wald (1943) demonstrated **D**-optimality property of the **LSD**s. (See also Nandi (1950)). Later, Ehrenfeld (1955) established **E**-optimality property of the **LSD**s. We have only added the detailed steps in connection with **D**-optimality for $v \geq 6$ in nonregular set-up and that too in the spirit of similar considerations developed by Kiefer (1975) in connection with **A**-optimality for all v in nonregular set-up[1]. As Kiefer observes, the point to be noted is that the above argument (for **D**-optimality) may break down for $v = 4$ and as a matter of fact, uptil now, no **D**-optimality result is available in this case. Instead, what has been demonstrated by Kiefer (1975) is, in fact, nonoptimality of the **GYD**s for $b = k$, $v = 4$ as regards **D**-optimality. We cite his counter-example with $b = k = 6$, $v = 4$.

GYD						design better than the **GYD**					
1	4	2	4	3	2	1	4	2	4	3	2
2	1	4	3	3	4	2	1	4	3	3	4
2	3	1	3	4	2	2	3	1	3	4	2
1	3	3	1	2	4	**4**	3	3	1	2	4
4	**1**	4	2	1	3	4	**2**	4	2	1	3
3	2	**1**	4	2	1	3	2	**3**	4	2	1

Kiefer has rightly remarked that this area of research is very incomplete due to the computational difficulties for designs not having simple structures as the above two designs. See Das and Dey (1988) for further results.

Remark 4.3.3. Cheng (1981) observed that analogous **A**- and **D**-optimality results also hold for the **PYD**s.

Cheng (1978) extended these results to the set-up of n-way heterogeneity and established universal optimality of what he termed *Youden Hyperrectangles* (**YHR**s) in the *regular* set-up. Further, in *non-regular* set-ups, he demonstrated **E**-optimality property of the **YHR**s and the **A**- and **D**-optimality of what he termed *Youden Hypercubes* (**YHC**s) for $n \geq 3$. The peculiar result that the **GYD**s (which are essentially 2-dimensional **YHC**s when $b = k$) are *not* **D**-optimal for $b = k$, $v = 4$ thus occurs only for $n = 2$. The **A**- and **D**-optimality of general **YHR**s have also been subsequently reported in Cheng (1980). We omit the discussion of these results.

---

[1]After the first draft of this monograph had been prepared, the authors understood that (the late) Professor Kiefer had presented the proof of $\Gamma_0(c_0+1) \geq 0$ in his paper "Optimal design theory in relation to combinatorial design" during a Conference in the Colorado State University in June, 1978 on "Optimum and Combinatorial design".
However, the authors intend to keep on record this proof as it simply indicates that there is some flexibility in getting into this result. The reader could as well develop his/her own arguments to get this result.

Mukhopadhyay and Mukhopadhyay (1981) considered a more general multiway set-up, allowing the set of experimental units to be the smallest possible for orthogonal estimation of all the parameters in the model. Optimality results parallel those of Cheng (1978) were established in Mukhopadhyay and Mukhopadhyay (1981) and a series of such optimal designs are presented in Bagchi and Mukhopadhyay (1983).

4. Optimality of Other Row-Column Designs

At the very outset we mention that the scope of availability of optimal designs, other than the **GYDs**, is limited. (See Jacroux (1986, 1987) for some illustrative examples of other E- and **MV**-optimal row-column designs.) Over the past few years, attempts have, however, been made to construct *highly efficient* designs. Again, precise results are, as yet, lacking. We attempt to put the few results available in a proper perspective to eventually produce some optimality results not hitherto found in the literature.

In subsection 4.1, we will present a detailed account of the bounds to the efficiency factor of a row-column design. In subsection 4.2, we present some optimality results followed by a discussion. Some examples of such optimal and/or highly efficient designs are given in subsection 4.3.

4.1 Efficiency Bounds for Row-Column Designs

The efficiency bounds analogous to those in a block design (viz., (3.3.5), (3.3.9) and (3.3.10)) can be constructed. See, for example, John and Eccleston (1986). We will *not* pursue this any further. Instead, we start with a study of the efficiency factor of a connected row-column design in relation to those of the two *component* block designs. We shall first observe that

$$E_{RC} \leq \min(E_R, E_C) \tag{4.4.1}$$

where E_{RC} is the efficiency factor of the row-column design and $E_R(E_C)$ is that of the associated block design with rows (columns) as blocks. To see this as also other inequality relations, let us consider the following **C**-matrices:

$$\mathbf{C} = \mathbf{r}^{\delta} - \frac{1}{b}\mathbf{M}\mathbf{M}' - \frac{1}{k}\mathbf{N}\mathbf{N}' + \frac{\mathbf{r}\mathbf{r}'}{bk} \tag{4.4.2a}$$

This is the **C**-matrix of the row-column design we start with.

$$\mathbf{C}_1 = \mathbf{r}^{\delta} - \frac{1}{b}\mathbf{M}\mathbf{M}' \tag{4.4.2b}$$

This is the **C**-matrix of the associated row design (assuming column effects to be all equal).

$$\mathbf{C}_2 = \mathbf{r}^{\delta} - \frac{1}{k}\mathbf{N}\mathbf{N}' \tag{4.4.2c}$$

This is the C-matrix of the associated column design (assuming row effects to be all equal).

$$\mathbf{C}_0 = \mathbf{r}^{\delta} - \frac{\mathbf{r}\mathbf{r}'}{bk} \tag{4.4.2d}$$

This is the C-matrix of the associated *Completely Randomized Design* (**CRD**) (assuming no differential row/column effects). It is clear that

$$\mathbf{C} = \mathbf{C}_1 + \mathbf{C}_2 - \mathbf{C}_0 \text{ and } \operatorname{rank}(\mathbf{C}_i) = v-1, \quad i = 0,1,2 \tag{4.4.3}$$

as the row-column design is assumed to be connected and $\mathbf{C}_i - \mathbf{C}$ can be shown to be nnd, $i = 0,1,2$.

It is also evident that $\mathbf{C}^+ - \mathbf{C}_1^+$ and $\mathbf{C}^+ - \mathbf{C}_2^+$ are both nnd. Here $\mathbf{A}^+$ denotes the Moore-Penrose (generalized) inverse of $\mathbf{A}$. We thus have $tr(\mathbf{C}^+) \geq tr(\mathbf{C}_i^+)$, $i = 1,2$. Since the efficiency factor of a design is inversely proportional to $tr(\mathbf{C}^+)$, the above inequality in (4.4.1) follows.

It is true that the above bound to E_{RC} is a weak one and it could be strengthened at least for an important subclass of designs. To develop such a bound, we first consider the notion of *basic contrasts* or *canonical contrasts* for a row-column design. As we have noted earlier, from a given row-column design, we may have two component block designs and also an associated **CRD** with the same replication numbers. For the subclass of designs to be studied here, there exist a complete set of $(v-1)$ treatment contrasts for which the estimates w.r.t. each of the above four designs (assuming for technicality, the underlying respective models to hold) are uncorrelated. To see this, let us define

$$\mathbf{A} = \mathbf{r}^{-\delta/2}\mathbf{C}\mathbf{r}^{-\delta/2}, \quad \mathbf{A}_i = \mathbf{r}^{-\delta/2}\mathbf{C}_i\mathbf{r}^{-\delta/2}, \quad i = 0,1,2.$$

It is easy to verify that $\mathbf{r}^{-\delta/2}\mathbf{A}^+\mathbf{r}^{-\delta/2}$ is a g-inverse of $\mathbf{C}$ and that $\mathbf{r}^{-\delta/2}\mathbf{A}_i^+\mathbf{r}^{-\delta/2}$ is a g-inverse of $\mathbf{C}_i$, $i = 0,1,2$. We note that $\mathbf{A}_0 = \mathbf{I} - \dfrac{\mathbf{r}^{-\delta/2}\mathbf{J}\mathbf{r}^{-\delta/2}}{bk}$. Further, we recall that $\mathbf{A}+\mathbf{A}_0 = \mathbf{A}_1+\mathbf{A}_2$. Since $\mathbf{r}^{-\delta/2}\mathbf{1}/\sqrt{bk}$ is an eigenvector of each of $\mathbf{A}$, $\mathbf{A}_0$, $\mathbf{A}_1$ and $\mathbf{A}_2$, whenever $\mathbf{A}_1$ and $\mathbf{A}_2$ *commute*, it is possible to find a common set of orthonormal eigenvectors simultaneously for $\mathbf{A}$, $\mathbf{A}_0$, $\mathbf{A}_1$ and $\mathbf{A}_2$. Hence, if we denote such a set by $\mathbf{p}_i (i = 1,2, \ldots, v)$ with $\mathbf{p}_v = \mathbf{r}^{-\delta/2}\mathbf{1}/\sqrt{bk}$, the $(v-1)$ treatment contrasts $\mathbf{l}_i'\tau$ with $\mathbf{l}_i = r^{-\delta/2}\mathbf{p}_i$ possess uncorrelated estimates w.r.t. each of these four designs. Also if λ_i, ϕ_i and θ_i, $1 \leq i \leq v-1$, denote respectively the eigenvalues of $\mathbf{A}$, $\mathbf{A}_1$ and $\mathbf{A}_2$ corresponding to $\mathbf{p}_i$ $(1 \leq i \leq v-1)$, then these are said to be the efficiencies in respective order of the row-column design, the row-design and the column-design w.r.t. the **CRD**, for the treatment contrast $\mathbf{l}_i'\tau$. The above efficiencies are also known as the *canonical efficiencies* and the treatment contrasts $\{\mathbf{l}_i'\tau, 1 \leq i \leq v-1\}$ are said to constitute a full set of orthonormal *canonical contrasts* or *basic contrasts*. We note incidentally

$$\text{(i) } 0<\lambda_i\leq\phi_i,\ \theta_i\leq 1,\ \text{(ii) } 1+\lambda_i = \phi_i+\theta_i,\ 1\leq i\leq v-1 \tag{4.4.4}$$

and, hence, that (iii) $\lambda_i\leq\phi_i\theta_i,\ 1\leq i\leq v-1$.

The above analysis is based on the assumption that $\mathbf{A}_1\mathbf{A}_2 = \mathbf{A}_2\mathbf{A}_1$. It is clear that this is equivalent to $\mathbf{MM'r^{-\delta}NN'}$ being a symmetric matrix. This will be referred to as the property of *commutativity* of the row-column design and the subclass of (connected) row-column designs (in a given set-up) for which this holds will be called the *commutative subclass*. The canonical efficiency factors for the first three designs in (4.4.2a), (4.4.2b), (4.4.2c) w.r.t. the one in (4.4.2d) are defined as the harmonic means of the λ_i's, ϕ_i's and θ_i's respectively and are denoted by $\mathcal{E}$, $\mathcal{E}_1$ and $\mathcal{E}_2$ in that order. Observing that $1/\lambda_i \geq 1/\phi_i + 1/\theta_i-1,\ 1\leq i\leq v-1$, we easily deduce that

$$\mathcal{E}^{-1}\geq \mathcal{E}_1^{-1}+\mathcal{E}_2^{-1}-1 \tag{4.4.5}$$

Further, equality holds iff

$$\lambda_i = \theta_i\phi_i,\ 1\leq i\leq v-1\ \text{ (as } \lambda_i = \theta_i+\phi_i-1 \text{ for all } i). \tag{4.4.6}$$

This condition is seen to be equivalent to $(1-\theta_i)(1-\phi_i) = 0,\ 1\leq i\leq v-1$ so that θ_i and/or $\phi_i = 1$ for every i. The above result was proved in Shah and Eccleston (1986) who also characterized the designs satisfying (4.4.6). It turns out that (4.4.6) holds iff $\mathbf{M'r^{-\delta}N} = \mathbf{J}$ in which case the row totals *adjusted* for treatments and the column totals *adjusted* for treatments are *mutually* uncorrelated. This property is called the property of *adjusted orthogonality* (Eccleston and Russell (1975)). It is easy to see that for a design with adjusted orthogonality, commutativity property obtains. When the design is equireplicate, $\mathcal{E}$ coincides with the *usual* efficiency factor E and, hence, for an *equireplicate commutative* design

$$E_{RC}^{-1} \geq E_R^{-1} + E_C^{-1} - 1. \tag{4.4.7}$$

Again, equality holds iff the design has adjusted orthogonality. The inequality (4.4.7) provides a non-trivial upper bound for the efficiency factor of a row-column design in the commutative subclass in terms of the efficiency factors of the two component designs. Thus, for two *given* component designs, if a design with adjusted orthogonality exists, it will possess the highest efficiency factor. This aspect of adjusted orthogonality will be exploited in the next subsection to establish optimality of certain row-column designs.

The inequality in (4.4.7) was first proved in Eccleston and McGilchrist (1985) where it was stated *without* the condition of commutativity. However, their proof does *not* seem to be valid for non-commutative designs and it is *not* clear if the inequality is true for designs of the non-commutative type.

4.2 Optimality of Some Classes of Adjusted Orthogonal Row-Column Designs

For given b, v, and k we shall initially consider the subclass (assumed to be non-null) of (connected) equireplicate commutative row-column designs. Consider an optimality functional $f(.)$ which is nonincreasing and convex, and call ψ_f the corresponding optimality criterion (which seeks to minimize $\psi_f = \sum_1^{v-1} f(x_i)/(v-1)$, x_i's being the non-zero eigenvalues of the C-matrix). We note that for a competing design, $r\lambda_i$, $r\phi_i$ and $r\theta_i$, $1 \le i \le v-1$, are the non-zero eigenvalues of $\mathbf{C}$, $\mathbf{C}_1$ and $\mathbf{C}_2$ respectively. In view of the convexity of f, it then follows that (recall (4.4.4))

$$f(r\lambda_i) \ge f(r\phi_i) + f(r\theta_i) - f(r), \quad 1 \le i \le v-1 \tag{4.4.8}$$

and, hence, for the ψ_f-criterion, we have

$$\psi_f(\mathbf{C}) \ge \psi_f(\mathbf{C}_1) + \psi_f(\mathbf{C}_2) - \psi_f(\mathbf{C}_0) \tag{4.4.9}$$

Since every competing design is commutative, it is clear that $r + r\lambda_i = r(\phi_i + \theta_i)$, $1 \le i \le v-1$. Further, if such a design is also adjusted orthogonal, then (4.4.6) holds. These together imply $(1-\theta_i)(1-\phi_i) = 0$ for every i, $1 \le i \le v-1$. Hence, recalling the definition of $\psi_f(\mathbf{C})$, we deduce that for an adjusted orthogonal equireplicate design

$$\psi_f(\mathbf{C}) = \psi_f(\mathbf{C}_1) + \psi_f(\mathbf{C}_2) - \psi_f(\mathbf{C}_0). \tag{4.4.10}$$

Observe that $\psi_f(\mathbf{C}_0) = f(r)$ is design-independent. Thus, if for a row-column design, the two component block designs are ψ_f-optimal within the corresponding classes of equireplicate block designs and if the row-column design is adjusted orthogonal, then it is ψ_f-optimal within the equireplicate commutative subclass. The class of optimality criteria considered here includes A- and D-optimality. This argument can be extended to any generalized criterion and to the class of Φ_p-criteria. In particular, it also holds for E-optimality criterion. Thus, essentially, we have enunciated the following

Theorem 4.4.1. For given (b,v,k), suppose there exists a row-column design which is equireplicate and adjusted orthogonal. Suppose further that each of its block-design components is ψ_f-optimal in the relevant equireplicate class. Then the row-column design is ψ_f-optimal in the subclass of all equireplicate commutative row-column designs in $\mathbf{D}(b,v,k)$.

Remark 4.4.1. It would be nice to extend this optimality result in the following directions. One would be to require optimality of at most one of the two component block designs and not both. The other would be to remove the restriction of equireplicability and/or commutativity of the competing designs. We have been able to accomplish both w.r.t. the E-optimality criterion. The precise result is given below.

Theorem 4.4.2. For given (b,v,k), suppose there exists a row-column design which is equireplicate and adjusted orthogonal. Suppose that one of its component block designs (say, the row-component, with the C-matrix $\mathbf{C}_1$ having roots $r\phi_i$, $1 \leq i \leq v-1$) is E-optimal. Whenever for the underlying $\mathbf{C}_1$ and $\mathbf{C}_2$ matrices, $\phi_{(1)}$ (the smallest ϕ_i) $\leq \theta_{(1)}$ (the smallest θ_i), the row-column design is E-optimal in the entire class $\mathbf{D}(b,v,k)$.

Proof. First observe that for any competing design

$$\mathbf{C} = \mathbf{C}_1 + \mathbf{C}_2 - \mathbf{C}_0 = \mathbf{C}_1 - (\mathbf{C}_0 - \mathbf{C}_2).$$

Hence, as $\mathbf{C}_0 - \mathbf{C}_2$ is nnd, minimum positive eigenvalue of $\mathbf{C} \leq$ minimum positive eigenvalue of $\mathbf{C}_1 \leq r\phi_{(1)}$ by the hypothesis of the above Theorem. Further, since the given row-column design is adjusted orthogonal and also $r\phi_{(1)} \leq r\theta_{(1)}$ is assumed to hold, it turns out that $r\phi_{(1)} =$ minimum positive eigenvalue of the C-matrix of the given row-column design which is thus proved to be E-optimal in the entire class.

Bagchi and Shah (1988) have recently proved that a row-column design which is adjusted orthogonal and for which the two component designs are **LBDs** is ϕ_f-optimal and Schur-optimal within the class of equireplicate designs. Their proof is based on the verification of weak upper majorization discussed in subsection 3.6 of Chapter One. Since the competing designs do *not* have to be commutative, designs of the above form, whenever available, have stronger optimality properties than those claimed in Theorem 4.4.1. Applications of the above result are given in the next subsection.

Remark 4.4.2. It seems difficult to relax the condition of Theorem 4.4.1 in general. Particularly, commutativity is seen to play a central role in the study of the C-matrices. Of course, the class of commutative designs is a very wide class. For example, if any of the component block designs is a **BBD**, then commutativity holds. On the other hand, we do *not* see a clear statistical interpretation of the property of commutativity on the part of a row-column design.

Remark 4.4.3. As regards adjusted orthogonality, it has a nice statistical interpretation as mentioned earlier. Further, it provides a tool to establish optimality of certain designs. However, as is clear from the hypothesis of Theorem 4.4.1, the component block designs need be optimal for a row-column design with adjusted orthogonality to be optimal. This feature may *not* always be achieved. In addition, the competing designs are restricted to the equireplicate commutative subclass. To illustrate this point, we consider a subclass of row-column designs with adjusted orthogonality (called α-designs) given in John and Eccleston (1986). For $b = 6$, $k = 4$ and $v = 12$, the *best* (i.e., most efficient) α-design has efficiency factor 0.627 and the design is shown below along with another design which is neither adjusted orthogonal nor commutative but has efficiency factor 0.648.

John and Eccleston (1986) Design						Alternative Design Shah and Puri (1986)					
12	4	8	3	7	11	7	3	2	10	6	5
1	5	9	12	4	8	3	8	1	6	11	4
2	6	10	5	9	1	2	1	9	5	4	12
3	7	11	10	2	6	12	10	11	9	7	8

Remark 4.4.4. Jacroux (1982) has furnished some results on **E**-optimality of certain row-column designs with $b>v>k$ obtained by adding some disjoint columns to a Youden Design. See also Jacroux (1987) for additional results on **E**-optimal and **MV**-optimal row-column designs.

4.3 Examples of Optimal/Highly Efficient Row-Column Designs

Below we discuss some classes of row-column designs.

(I) Anderson and Eccleston (1985) start with a symmetric **BIBD** with parameters $(b^*, v^*, r^*, k^*, \lambda^*)$ whose incidence matrix $\mathbf{N}^*$ can be written as

$$\mathbf{N}^* = \begin{pmatrix} \mathbf{A} & \mathbf{0} \\ \mathbf{B} & \mathbf{1} \end{pmatrix}$$

Andersen and Eccleston (1985) have referred to some algorithms which can be used to construct a row-column design with parameters $b = v^*-k^*$, $k = k^*$, $v = v^*-1$ and having the row-treatment and column-treatment incidence matrices as $\mathbf{M} = \mathbf{J}-\mathbf{B}'$ and $\mathbf{N} = \mathbf{A}'$. It is easy to see that the row-column design so constructed is equireplicate and adjusted orthogonal. Further, each component block design is an **LBD**. Hence, by the result of Bagchi and Shah (1988) these designs are ψ_f-optimal in the class of equireplicate designs. Also, by Theorem 4.4.2 these designs are **E**-optimal in the entire class $\mathbf{D}(b,v,k)$.

(II) Agrawal (1966) gave a class of designs in $4\lambda+3$ rows, $4\lambda+4$ columns and $2(4\lambda+3)$ treatments where $4\lambda+3$ is a prime. These designs (described in Method 3.5 of the above cited paper) can be seen to be adjusted orthogonal. Further, it can be verified that for these designs, each component design is an **LBD**. Hence these designs have the same optimality properties as the designs given by Anderson and Eccleston described above.

(III) Russell (1980) gives examples of a series of row-column designs with parameters $b>v>k$ whose column-component block designs are **BIBDs**. As to the row-component block designs, they are generalized binary and duals of **BBDs**. He has tabulated the ratio E_{RC}/E_C for all the designs and these are found to be quite high. These designs are not, of course, adjusted orthogonal. We are thus unable to establish their optimality through the use of Theorem 4.4.1. Clearly, new tools are needed to examine the optimality aspects of such designs.

REFERENCES

Agrawal, H.L. (1966). Some systematic methods of construction of designs for two-way elimination of heterogeneity. *Cal. Statist. Assoc. Bull, 15*, 93-108.

Anderson, D.A. and Eccleston, J.A. (1985). On the construction of a class of efficient row-column designs. *J. Statist. Planning and Inference, 11*, 131-134.

Ash, A. (1981). Generalized Youden designs: Construction and Tables. *J. Statist. Planning and Inference, 5*, 1-25.

Bagchi, S. and Mukhopadhyay, A.C. (1983). Construction of balanced Youden hyperrectangles. *Tech. Report*, Computer Science Unit, Indian Statistical Institute, Calcutta.

Bagchi, S. and Shah, K.R. (1988). On the optimality of a class of row-column designs. *J. Statist. Planning and Inference*. To appear (1988-89).

Cheng, C.S. (1978). Optimal designs for the elimination of multi-way heterogeneity. *Ann. Statist., 6*, 1262-1272.

Cheng, C.S. (1980). Optimal designs for the elimination of multi-way heterogeneity II (Preliminary report): *IMS Bull. 9*, April 1980 (No. 80t-57).

Cheng, C.S. (1981). Optimality and construction of pseudo-Youden designs. *Ann. Statist., 9*, 200-205.

Das, A. and Dey, A. (1988). On universal optimality and non-optimality of row-column designs. Submitted to *Jour. Statist. Planning and Inference*.

Eccleston, J.A. and Russell, K.G. (1975). Connectedness and orthogonality in multi-factor designs. *Biometrika, 62*, 341-345.

Eccleston, J.A. and McGilchrist, C.A. (1985). Algebra of a row-column design. *J. Statist. Planning and Inference, 12*, 305-310.

Ehrenfeld, S. (1955). On the efficiency of experimental designs. *Ann. Math. Statist., 26*, 247-255.

Jacroux, M. (1982). Some E-optimal designs for the one-way and two-way elimination of heterogeneity. *J.R. Statist. Soc. B, 44*, 253-261.

Jacroux, M. (1986). Some E-optimal row-column designs. *Sankhyā (B), 48*, 31-39.

Jacroux, M. (1987). Some E- and MV-optimal row-column designs having an equal number of rows and columns. *Metrika, 34*, 361-381.

John, J.A. and Eccleston, J.A. (1986). Row-column α-designs. *Biometrika, 73*, 301-306.

Kiefer, J. (1958). On the nonrandomized optimality and randomized nonoptimality of symmetrical designs. *Ann. Math. Statist.*, *29*, 675-699.

Kiefer, J. (1959). Optimum experimental designs. *J.R. Statist. Soc., Ser. B, 21*, 272-319.

Kiefer, J. (1971). The role of symmetry and approximation in exact design optimality. In *Statistical decision theory and related topics.* (Ed. S.S. Gupta and James Yackel) Academic Press. New York and London, 109-118.

Kiefer, J. (1975). Construction and optimality of generalized Youden designs. *A Survey of Statistical Designs and Linear Models.* (J.N. Srivastava, ed.) North Holland Publ. Co., Amsterdam 333-353.

Mukhopadhyay, A.C. and Mukhopadhyay, S. (1981). Optimality in a balanced incomplete multiway heterogeneity set-up. In *Statistics: Applications and New Directions.* Proceedings of the Indian Statistical Institute Golden Jubilee International Conference, 466-477.

Nandi, H.K. (1950). On the efficiency of experimental designs. *Cal. Stat. Assoc. Bull.*, *3*, 161-171.

Raghavarao, D. (1971). *Constructions and combinatorial problems in designs of experiments.* John Wiley, New York.

Ruiz, F. and Seiden, E. (1974). On construction of some families of generalized Youden designs. *Ann. Statist.*, *2*, 503-519.

Russell, K.G. (1980). Further results on the connectedness and optimality of designs of type $O: XB$. *Commun. Statist. (A) Theor. Meth.*, *9*, 439-447.

Saharay, R. (1986). Optimal designs under a certain class of non-orthogonal row-column structure. *Sankhyā (B)*, *48*, 44-67.

Shah, K.R. and Eccleston, J.A. (1986). On some aspects of row-column designs. *J. Statist. Planning and Inference*, *15*, 85-95.

Shah, K.R. and Puri, P.D. (1986). Commutative row-column designs. (Unpublished).

Sinha, B.K. (1980). Some further combinatorial and constructional aspects of generalized Youden designs. *Combinatorics and graph theory* (Calcutta 1980), pp. 488-493. Lecture Notes in Math, 885, Sprigner, Berlin-New York 1981. 83g.62118.

Wald, A. (1943). On the efficient design of statistical investigations. *Ann. Math. Statist.*, *14*, 134-140.

CHAPTER FIVE

MIXED EFFECTS MODELS

1. Introduction

In most design set-ups such as block designs or row-column designs classification effects such as block effects or row (column) effects are regarded fixed. When these are also considered random variables, we have one or more additional sources of information for estimating treatment effect parameters. Such models are known as mixed effects models. In the experimental designs these were first introduced by Yates (1939).

In this Chapter, we shall discuss the optimality of designs when the analysis is based on an appropriate mixed effects model. In most cases, we shall attempt to show that a design which is optimal under the fixed effects model continues to remain so under the mixed effects model. These results are non-trivial because the comparative position of two given designs w.r.t. a given criterion such as **A**- or **D**-optimality may *not* remain the same under the two models. This will be made more explicit with some examples.

In section 2 we shall discuss the optimality results for the block designs. The results for the row-column designs will be discussed in section 3. It is important to recognize that the analysis of a mixed effects model is based on the estimation of the variance components and the estimates of the treatment effects are affected by the errors in the estimation of these variance components. For the purpose of optimality, we shall ignore these errors i.e. we shall regard the variance components as known up to a scalar multiplier.

Our primary interest will be in optimality results which hold for *all* values of the unknown variance components. Thus, optimality w.r.t. the fixed effects model is indeed an apriori necessary condition for optimality w.r.t. the mixed effects model. For this reason, even though the designs which are disconnected in the fixed effects model *are* connected in the mixed effects model these could *not* be optimal for all values of the variance components and hence need not be considered.

Optimality of block designs under the mixed effects model was first considered by Sinha (1980). In a series of papers, Mukhopadhyay (1981, 1984, 1987) and Bagchi (1987) continued this work and established optimality of certain designs. Khatri and Shah (1981) and Bhattacharya and Shah (1984) also obtained results which overlap with those of Mukhopadhyay (1984). Our treatment here is mainly based on the work of these authors. Work in this area is still in early stages and many important problems yet remain unresolved.

2. Optimality Aspects of Block Designs Under a Mixed Effects Model

In this section we shall consider designs in $\mathbf{D}(b,v,k)$ with $k<v$. It would be convenient to write the model giving the $bk\times 1$ vector of observations $\mathbf{Y}$ (taken across the blocks) as

$$\mathbf{Y} = \mathbf{X}_1\tau + \mathbf{X}_2\beta + \epsilon \tag{5.2.1}$$

where $\tau(v\times1)$ denotes the vector of treatment effects parameters, $\beta(b\times1)$ denotes the vector of block effects parameters assumed to be random, $\epsilon(bk\times1)$ denotes the vector of error variables while $\mathbf{X}_1(bk\times v)$ and $\mathbf{X}_2(bk\times b)$ are matrices of coefficients which are determined by the design. It is assumed that β and ϵ are independently distributed with

$$\epsilon \sim N(0,\sigma^2\mathbf{I}_{bk}),\ \beta \sim N(0,\sigma_\beta^2\mathbf{I}_b) \tag{5.2.2}$$

where σ^2 and σ_β^2 are unknown variance parameters. We note that $\mathbf{X}_1'\mathbf{X}_2$ equals $\mathbf{N}$, the incidence matrix of the design.

Rao (1947) showed that the matrix of coefficients in the normal equations for estimating the treatment contrasts is given by

$$\overline{\mathbf{C}} = (\mathbf{r}^\delta - \mathbf{NN}'/k) + \frac{1}{\rho}(\mathbf{NN}'/k - \mathbf{r}^\delta\mathbf{J}_{vv}\mathbf{r}^\delta/bk) \tag{5.2.3}$$

where $\rho = (\sigma^2+k\sigma_\beta^2)/\sigma^2$. In practice, ρ is usually estimated from the data. However, for the present study we shall regard it as known. It would sometimes be convenient to write $\overline{\mathbf{C}}$ in the following alternative form (due to Khatri and Shah (1981))

$$\overline{\mathbf{C}} = \mathbf{H}_1(\mathbf{H}_1'\mathbf{A}^{-1}\mathbf{H}_1)^{-1}\mathbf{H}_1' \tag{5.2.4}$$

where $\mathbf{A} = \mathbf{r}^\delta - a\mathbf{NN}'$, $a = (\rho-1)/\rho k$ and $\mathbf{H}_1$ is a $v\times(v-1)$ matrix such that $\mathbf{H}_1'\mathbf{H}_1 = \mathbf{I}_{v-1}$ and $\mathbf{H}_1\mathbf{H}_1' = \mathbf{B} = \mathbf{I}_v - \mathbf{J}_{vv}/v$.

It is easy to verify that rank $(\overline{\mathbf{C}}) = v-1$. The optimality results will be in terms of the matrix $\overline{\mathbf{C}}$. We are especially interested in the situation where the optimal design does not depend upon the value of ρ.

It may be noted from (5.2.3) that $tr(\overline{\mathbf{C}})$ is maximized by a design iff it satisfies (i) $|n_{ij}-n_{i'j'}|\le1$ and (ii) $|r_i-r_{i'}|\le1$ for $i,\ i'=1,2,\ldots,v$ and $j,\ j'=1,2,\ldots,b$. (Vide Jacroux and Seeley (1980)).

If the class $\mathbf{D}(b,v,k)$ contains a **BBD** then for that design $tr(\overline{\mathbf{C}})$ is maximum. Further, $\overline{\mathbf{C}}$ is completely symmetric. Hence by Proposition 1 (Restated) given in subsection 3.2 of Chapter One, the **BBD** is universally optimal in the extended sense w.r.t. the mixed effects model.

We shall now consider the analogue of ψ_f optimality w.r.t. the mixed effects model. Let $\bar{x}_i$, $i=1,2,\ldots,v-1$ denote the non-zero eigenvalues of $\overline{\mathbf{C}}$. Further, let f be a real-valued function which is strictly decreasing, strictly convex with $f'''<0$ and $f(0)=\infty$. A design is said to be $\overline{\psi}_f$ optimal if it minimizes $\psi_f(\overline{\mathbf{C}}) = \sum_{i=1}^{v-1} f(\bar{x}_i)$ among all designs in $\mathbf{D}(b,v,k)$. Similarly, a design will be said to be $\overline{\mathbf{A}}$-, $\overline{\mathbf{D}}$- or $\overline{\mathbf{E}}$-optimal if it minimizes $\sum\bar{x}_i^{-1} = tr(\mathbf{A}^-\mathbf{B})$, $\Pi\bar{x}_i^{-1} = (bk/v\rho)/|\mathbf{A}|$, or $\max(\bar{x}_i^{-1})$ respectively.

Finally, a design will be said to be $(\overline{\mathbf{M},\mathbf{S}})$-optimal if it minimizes $tr(\overline{\mathbf{C}}^2)$ among all designs which maximize $tr(\overline{\mathbf{C}})$.

We shall first establish the $\overline{\psi}_f$-optimality of a **GDD** with $m=2$ and $\lambda_2=\lambda_1+1$. To this end, we define $\overline{P}^2=tr(\overline{\mathbf{C}}^2)-(tr(\overline{\mathbf{C}}))^2/(v-1)$ and note that if a design maximizes $tr(\overline{\mathbf{C}})$ and maximizes $tr(\overline{\mathbf{C}})-\delta\overline{P}$ where $\delta^2=(v-1)/(v-2)$, by Theorem 2.3.2 of Chapter Two, it is $\overline{\psi}_f$-optimal.

By the results of Jacroux and Seely (1980), it can be shown that any **GDD** has the maximum value for $tr(\overline{\mathbf{C}})$. Further, it is easy to verify that

$$tr(\overline{\mathbf{C}}^2)=a^2k^2tr(\mathbf{C}^2)+2ak(1-ak)tr(\mathbf{C}_0\mathbf{C})+(1-ak)^2tr(\mathbf{C}_0^2)$$

where $\mathbf{C}_0=\mathbf{r}^\delta-\mathbf{r}^\delta\mathbf{J}_{vv}\mathbf{r}^\delta/bk$. From this, one can deduce that if

$$tr(\mathbf{C}\mathbf{C}_0)\geq(tr\ \mathbf{C}_0)(tr\ \mathbf{C})/(v-1) \tag{5.2.5}$$

then

$$\overline{P}^2\geq a^2k^2P^2$$

where, as before, $P^2=tr(\mathbf{C}^2)-(tr(\mathbf{C}))^2/(v-1)$.

We note that for any equireplicate design, $\overline{P}^2=a^2k^2P^2$. Denoting the expressions $\mathbf{C}$, $\overline{\mathbf{C}}$, P and $\overline{P}$ for the above **GDD** by $\mathbf{C}_*$, $\overline{\mathbf{C}}_*$, P_* and $\overline{P}_*$ respectively, for any design satisfying (5.2.5), we have

$$\overline{P}-\overline{P}_*\geq ak(P-P_*).$$

This gives us

$$\{tr(\overline{\mathbf{C}})-\delta\overline{P}\}-\{tr(\overline{\mathbf{C}}_*)-\delta\overline{P}_*\}\leq ak\{tr(\mathbf{C})-\delta P-(tr\ \mathbf{C}_*-\delta P_*)\}.$$

In subsection 3.3 of Chapter Two we had seen that a **GDD** with $m=2$ and $\lambda_2=\lambda_1+1$ maximizes $tr\ \mathbf{C}-\delta P$. Since $a>0$, it is clear that $tr\ \overline{\mathbf{C}}-\delta\overline{P}\leq tr\overline{\mathbf{C}}_*-\delta\overline{P}_*$.

The above argument establishes the $\overline{\psi}_f$-optimality of a **GDD** with $m=2$ and $\lambda_2=\lambda_1+1$ among the class of designs which satisfy (5.2.5). Bhattacharya and Shah (1984) have shown that (5.2.5) holds if the design is (i) binary or (ii) equireplicate or (iii) variance-balanced i.e. $\overline{\mathbf{C}}$ is completely symmetric or (iv) efficiency balanced i.e. $\overline{\mathbf{C}}$ is proportional to $\mathbf{C}_0$.

The result is much harder to establish without this condition. This has been recently done by Mukhopadhyay (1987) by showing that a design which does not satisfy (5.2.5) cannot improve upon the **GDD** under consideration w.r.t. $\overline{\psi}_f$-optimality.

We shall now establish the $\overline{\psi}_f$-optimality of the Linked Block Design (**LBD**) among an appropriate class of *equireplicate* designs assuming only that f is non-increasing and convex. If d^* is an **LBD**, the non-zero eigenvalues of $\overline{\mathbf{C}}_{d^*}$ are given by

$$\bar{x}_{i^*} = \begin{cases} r & \text{for } i = 1,2, \ldots, v-b \\ r-a\mu & \text{for } i = v-b+1, \ldots, v-1 \end{cases}$$

where $\mu = k(b-r)/(b-1)$. If d is any other equireplicate design in $\mathbf{D}(b,v,k)$ then the corresponding non-zero eigenvalues of $\overline{\mathbf{C}}_d$ are

$$\bar{x}_i = \begin{cases} r & \text{for } i = 1,2, \ldots, v-b \\ r-a\mu_{i-(v-b)} & \text{for } i = v-b+1, \ldots, v-1 \end{cases}$$

where $\mu_1,\mu_2, \ldots, \mu_{b-1}$, $\mu_b = rk$ are the eigenvalues of $\mathbf{N'N}$ for the design d. Hence, we get

$$\psi_f(\overline{\mathbf{C}}_{d^*}) - \psi_f(\overline{\mathbf{C}}_d) = (b-1)f(r-a\mu) - \sum_{i=1}^{b-1} f(r-a\mu_i).$$

Since $r-a\mu \geq \frac{1}{(b-1)}\sum(r-a\mu_i)$, we have

$$f(r-a\mu) \leq f(\frac{1}{b-1}\sum(r-a\mu_i)) \leq \frac{1}{b-1}\sum f(r-a\mu_i)$$

and hence $\psi_f(\overline{\mathbf{C}}_{d^*}) \leq \psi_f(\overline{\mathbf{C}}_d)$. Thus, an **LBD** is $\overline{\psi}_f$-optimal within the sub-class of equireplicate designs in $\mathbf{D}(b,v,k)$.

We shall now discuss some specific optimality aspects for designs under this mixed effects model.

(i) $\overline{(\mathbf{M,S})}$-optimality: We have already stated the necessary and sufficient conditions for maximizing $tr(\overline{\mathbf{C}})$. This requires $|r_i - r_{i'}| \leq 1$ for $i, i' = 1,2, \ldots, v$. Let us assume that $r_i = \alpha = [bk/v]$ for $i = 1,2, \ldots, t$ and $r_i = \alpha+1$ for $i = t+1, \ldots, v$. It turns out that $tr(\overline{\mathbf{C}}^2)$ is minimum iff

$$b(\rho-1)\sum_{i>j}\lambda_{ij}^2 + 2\sum_{i>j=t+1}^{v} \lambda_{ij} \text{ is minimum}$$

subject to the conditions

$$\sum_{j=1}^{v}\lambda_{ij} = \begin{cases} \alpha k & \text{for } i \leq t \\ (\alpha+1)k & \text{for } i > t \end{cases}$$

where $\lambda_{ij} = \sum_l n_{il}n_{jl}$. Thus, if a design minimizes *each* of $\sum_{i>j}\lambda_{ij}^2$ and $\sum_{i>j=t+1}^{v} \lambda_{ij}$ among all designs which maximize $tr(\overline{\mathbf{C}})$, it is $\overline{(\mathbf{M,S})}$-optimal. Hence, if $t = 1$, $v-1$ or v, a design which minimizes $\sum_{i>j}\lambda_{ij}^2$ among all designs which maximize $tr(\overline{\mathbf{C}})$, is $\overline{(\mathbf{M,S})}$ optimal for *all* values of ρ.

It should be noted that the comparative position of two designs d_1 and d_2 w.r.t. the $\overline{(\mathbf{M,S})}$-optimality criterion may well depend upon the value of ρ. To see this, we consider three designs d_1, d_2 and d_3 given below.

	Blocks				
	1	2	3	4	5
	1	2	3	1	4
d_1	2	4	4	5	5
	3	5	6	6	6

	Blocks				
	1	2	3	4	5
	2	3	2	1	1
d_2	3	4	5	6	4
	4	6	6	5	5

	Blocks				
	1	2	3	4	5
	1	1	2	2	3
d_3	5	5	3	4	4
	4	6	5	6	6

For each design $r_1 = r_2 = r_3 = 2$ and $r_4 = r_5 = r_6 = 3$. Further, $tr(\overline{\mathbf{C}}_{d_1}) = tr(\overline{\mathbf{C}}_{d_2}) = tr(\overline{\mathbf{C}}_{d_3})$. It turns out that $tr(\overline{\mathbf{C}}_{d_2}^2) \lesseqgtr (tr\ \overline{\mathbf{C}}_{d_1}^2)$ if $\rho \lesseqgtr \frac{6}{5}$. Also, $tr(\overline{\mathbf{C}}_{d_3}^2) < tr(\overline{\mathbf{C}}_{d_1}^2)$ or $tr(\overline{\mathbf{C}}_{d_2}^2)$ for *all* values of ρ. Thus neither d_1 nor d_2 is uniformly better than the other but d_3 is uniformly better than both d_1 and d_2 w.r.t. $\overline{(\mathbf{M,S})}$ optimality criterion.

(ii) $\overline{\mathbf{A}}$ and $\overline{\mathbf{D}}$-optimalities: For the rest of this section, we shall confine ourselves to equireplicate designs only. Let E denote the efficiency factor of a block design. It is given by $E = (v-1)/r(\sum x_i^{-1})$ where as before, x_i's are the non-zero eigenvalues of the $\mathbf{C}$-matrix. Let $\overline{E} = (v-1)/r(\sum \overline{x}_i^{-1})$ denote the corresponding expression for the efficiency factor under a mixed effects model when ρ is known. Further, let E^* denote the corresponding expression when the estimates of the treatment effects are based upon the value of ρ estimated from the data. It would be reasonable to expect that

$$0<E\leq E^*\leq \overline{E}\leq 1 \tag{5.2.6}$$

for all ρ provided that a *suitable* estimate of ρ is used. This has been proved in Khatri and Shah (1981) for equireplicate designs.

We now consider comparison of two designs w.r.t. $\overline{\mathbf{A}}$- or $\overline{\mathbf{D}}$-optimality criteria. For equireplicate designs, the eigenvalues of $\overline{\mathbf{C}}$ and $\mathbf{C}$ have the relation $\overline{x}_i = (r+\rho_1 x_i)/\rho$ where $\rho_1 = \rho-1$.

We now consider designs d and d_1 and suppose that $\mathbf{C}_d$ has p distinct non-zero eigenvalues $z_1, z_2, \ldots, z_p$ with multiplicities $\alpha_1, \alpha_2, \ldots, \alpha_p$ and that $\mathbf{C}_{d_1}$ has q distinct eigenvalues $y_1, y_2, \ldots, y_q$ with multiplicities $\beta_1, \beta_2, \ldots, \beta_q$ respectively. We may write $tr\ \overline{\mathbf{C}}_d^+ - tr\ \overline{\mathbf{C}}_{d_1}^+$ as

$$h = tr\,\overline{\mathbf{C}}_d^+ - tr\,\overline{\mathbf{C}}_{d_1}^+ = \sum_{i=1}^{p} \frac{\alpha_i \rho}{r+\rho_1 z_i} - \sum_{j=1}^{q} \frac{\beta_j \rho}{r+\rho_1 y_j}. \tag{5.2.7}$$

This can be written in an alternative form as

$$h\{\prod_{i=1}^{p}(r+\rho_1 z_i)\}\{\prod_{j=1}^{q}(r+\rho_1 y_j)\} = \rho \sum_{l=1}^{p+q-1} h_l r^{p+q-l-1} \rho_1^l \tag{5.2.8}$$

where the h_l's are functions of z_i's, y_j's, α_i's and β_j's. It can be verified that

$$h_1 = tr(\mathbf{C}_{d_1}) - tr(\mathbf{C}_d)$$

$$h_2 = (tr(\mathbf{C}_d^2) - tr(\mathbf{C}_{d_1}^2)) + h_1(\sum_{i=1}^{p} z_i + \sum_{j=1}^{q} y_j)$$

and

$$h_{p+q-1} = \{\sum_{i=1}^{p} \frac{\alpha_i}{z_i} - \sum_{j=1}^{q} \frac{\beta_j}{y_j}\} \prod_{i=1}^{p} z_i \prod_{j=1}^{q} y_j.$$

The design d is better than the design d_1 w.r.t. the $\overline{\mathbf{A}}$-optimality criterion in the mixed effects model if $tr(\overline{\mathbf{C}}_d)<tr(\overline{\mathbf{C}}_{d_1})$ or equivalently if $h<0$. When ρ_1 is very large, the sign of h is determined by that of h_{p+q-1}. Similarly, when ρ_1 is very small, the sign of h_1 determines that of h so that $h<0$ if $tr(\mathbf{C}_d)>tr(\mathbf{C}_{d_1})$. If these traces are equal, $h<0$ for very small values of ρ_1 if $tr(\mathbf{C}_d^2)<tr(\mathbf{C}_{d_1}^2)$. When $p+q = 4$, d is better than d_1 *uniformly* in ρ if each of h_1, h_2 and h_3 is negative. We note that h_3 is negative if d is better than d_1 w.r.t. the $\mathbf{A}$-optimality criterion whereas h_1 and h_2 are negative if d is better than d_1 w.r.t. $\overline{(\mathbf{M,S})}$-optimality criterion. Thus, when $p+q = 4$, $\mathbf{A}$-optimality *and* $\overline{(\mathbf{M,S})}$-optimality imply $\overline{\mathbf{A}}$-optimality. In particular, if d and d_1 are both **PBIB** designs with two associate classes, we have $p+q = 4$ and the above case could arise. When $p+q>4$, no definite conclusions can be drawn regarding d being uniformly better than d_1 w.r.t. the $\overline{\mathbf{A}}$-optimality criterion.

Finally, to compare designs d and d_1 w.r.t. the $\overline{\mathbf{D}}$-optimality criterion, we note that the difference between the products of the non-zero eigenvalues of $\rho\overline{\mathbf{C}}_d$ and $\rho\overline{\mathbf{C}}_{d_1}$ can be expressed as

$$\sum_{j=1}^{v-1} (t_j - t_j^{(1)}) r^{v-1} (\rho_1/r)^j$$

where $t_j = tr_j(\mathbf{C}_d)$ and $t_j^{(1)} = tr_j(\mathbf{C}_{d_1})$. Here $tr_j(\mathbf{A})$ denotes the sum of the principal minors of order j for a matrix $\mathbf{A}$. Thus, if $t_{v-1}>t_{v-1}^{(1)}$, i.e. if d is better than d_1

w.r.t. **D**-optimality criterion, it will be better than d_1 w.r.t. the $\bar{\mathbf{D}}$-optimality criterion also for large values of ρ_1. For very small values of ρ_1 the difference is dominated by $t_1 - t_1^{(1)} = tr(\mathbf{C}_d) - tr(\mathbf{C}_{d_1})$. When $tr(\mathbf{C}_d) = tr(\mathbf{C}_{d_1})$, this is dominated by $t_2 - t_2^{(1)} = \{tr(\mathbf{C}_d^2) - tr(\mathbf{C}_{d_1}^2)\}/2$ for very small values of ρ_1. However, no conclusions can be reached for all values of ρ.

(iii) $\bar{\mathbf{E}}$-optimality: Recently, Bagchi (1987) demonstrated $\bar{\mathbf{E}}$-optimality of the following designs within the *entire* class $\mathbf{D}(b,v,k)$ of blocks designs:
(a) A **GDD** with $\lambda_2 = \lambda_1+1$
(b) A **GDD** with $\lambda_1 = \lambda_2+1$ and group size 2
(c) An **LBD**
(d) The dual of any design in (a) above
(e) The dual of any design in (b) above.

It can be seen that $\bar{\mathbf{E}}$-optimality is equivalent to **E**-optimality if we restrict to the subclass of equireplicate designs. The proofs of optimality of the above designs within the entire class $\mathbf{D}(b,v,k)$ are highly involved and we refer to Bagchi (1987) for a detailed account of the same.

3. Optimality of GYDs Under a Mixed Effects Model

We shall now discuss the optimality aspects of row-column designs under a mixed effects model. In this situation one may consider the model given by

$$\mathbf{Y} = \mathbf{X}_1\tau + \mathbf{X}_2\alpha + \mathbf{X}_3\beta + \epsilon \tag{5.3.1}$$

where $\mathbf{Y}(bk\times1)$ is the vector of observations, $\tau(v\times1)$ is the vector of treatment effect parameters, $\alpha(k\times1)$ is the vector of row effects, $\beta(b\times1)$ is the vector of column effects and $\epsilon(bk\times1)$ is the vector of observational errors. Here $\mathbf{X}_1,\mathbf{X}_2,\mathbf{X}_3$ are matrices of appropriate orders determined by the design. We note that $\mathbf{X}_1'\mathbf{X}_2 = \mathbf{M}$, $\mathbf{X}_1'\mathbf{X}_3 = \mathbf{N}$ and $\mathbf{X}_2'\mathbf{X}_3 = \mathbf{J}_{kb}$ where $\mathbf{M}$ and $\mathbf{N}$ are the incidence matrices of the row design and the column design respectively. In the mixed effects model considered here, we regard τ as fixed whereas α, β and ϵ are assumed to be independently normally distributed with

$$\alpha \sim N(0,\sigma_\alpha^2\mathbf{I}_k),\ \beta \sim N(0,\sigma_\beta^2\mathbf{I}_b) \text{ and } \epsilon \sim N(0,\sigma^2\mathbf{I}_{bk}). \tag{5.3.2}$$

The matrix of coefficients for the normal equations for estimating τ, i.e. the information matrix turns out to be

$$\bar{\mathbf{C}} = \mathbf{C} + \frac{1}{\rho_1}(\mathbf{C}_0-\mathbf{C}_1) + \frac{1}{\rho_2}(\mathbf{C}_0-\mathbf{C}_2) \tag{5.3.3}$$

where as defined earlier, $\mathbf{C}$, $\mathbf{C}_1$, $\mathbf{C}_2$ and $\mathbf{C}_0$ are the **C**-matrices for the row-column design, the block design with rows as blocks, the block design with columns as blocks and the **CRD** respectively. Here, $\rho_1 = \sigma^2 + b\sigma_\alpha^2$ and $\rho_2 = \sigma^2 + k\sigma_\beta^2$. For derivation of this, one may refer to Roy and Shah (1961).

Only available results on the optimality of row-column designs w.r.t. the above mixed model relate to the **GYDs**. The treatment here is due to Mukhopadhyay (1981).

We shall first establish universal optimality of the regular **GYDs** under this mixed effects model.

Since $\mathbf{C} + \mathbf{C}_0 = \mathbf{C}_1 + \mathbf{C}_2$ we may write

$$\overline{\mathbf{C}} = (1-\frac{1}{\rho_1})\mathbf{C} + (\frac{1}{\rho_1} - \frac{1}{\rho_2})\mathbf{C}_2 + \frac{1}{\rho_2}\mathbf{C}_0. \tag{5.3.4}$$

We note that $\rho_1 \geq 1$ and $\rho_2 \geq 1$. Further, a regular **GYD** maximizes each of $tr(\mathbf{C})$, $tr(\mathbf{C}_2)$ and $tr(\mathbf{C}_0)$. Thus when $\rho_1 \leq \rho_2$, a regular **GYD** maximizes $tr(\overline{\mathbf{C}})$. Similar argument holds when $\rho_1 \geq \rho_2$. Also each of $\mathbf{C}$, $\mathbf{C}_2$ and $\mathbf{C}_0$ is completely symmetric and hence $\overline{\mathbf{C}}$ is completely symmetric. Thus, by Proposition 1 (Restated) of Chapter One, a regular **GYD** is universally optimal in the usual and in the extended sense. Next, we note that we can write $\overline{\mathbf{C}}$ as

$$\overline{\mathbf{C}} = (1-\frac{1}{\rho_1})\mathbf{C}_1 + (1-\frac{1}{\rho_2})\mathbf{C}_2 + (\frac{1}{\rho_1} + \frac{1}{\rho_2} - 1)\mathbf{C}_0. \tag{5.3.5}$$

We note that a **GYD** maximizes each of $tr(\mathbf{C}_1)$, $tr(\mathbf{C}_2)$ and $tr(\mathbf{C}_0)$. Thus, when $(1/\rho_1) + (1/\rho_2) \geq 1$ i.e. when $(\rho_1-1)(\rho_2-1) \leq 1$, $\overline{\mathbf{C}}$ for a **GYD** (regular or non-regular) has maximum trace and is completely symmetric. Thus a non-regular **GYD** is also universally optimal when $(\rho_1-1)(\rho_2-1) \leq 1$.

We shall now turn to the non-regular **GYDs** when $(\rho_1-1)(\rho_2-1) > 1$. We shall use the following notation which builds upon the notation used in section 3 of Chapter Four for the fixed effects model. We shall denote the h-th diagonal element of $\mathbf{C}(\mathbf{C}_2)$ by $C_{hh}(C_{hh}^{(2)})$. As noted earlier $\mathbf{C}_2$ is the $\mathbf{C}$-matrix for the block design with columns as blocks. We now define

$$\phi(r_h) = \max C_{hh} \text{ subject to } \sum_j n_{hj} = r_h = \sum_l m_{hl}$$

$$\phi^{(2)}(r_h) = \max C_{hh}^{(2)} \text{ subject to } \sum_j n_{hj} = r_h$$

$$\phi^{(0)}(r_h) = r_h - r_h^2/bk$$

$$\overline{\phi}(r_h) = (1-\frac{1}{\rho_1})\phi(r_h) + (\frac{1}{\rho_1} - \frac{1}{\rho_2})\phi^{(2)}(r_h) + \frac{1}{\rho_2}\phi^{(0)}(r_h)$$

$$g(r_h) = bk\phi(r_h), \quad g^{(2)}(r_h) = bk\phi^{(2)}(r_h), \quad \overline{g}(r_h) = bk\overline{\phi}(r_h).$$

An analysis similar to the one in section 3 of Chapter Four indicates that $\bar{x}_{(1)}$, the smallest non-zero eigenvalue of $\overline{\mathbf{C}}$ satisfies

$$\bar{x}_1 \le \frac{v}{v-1}\min_h \overline{C}_{hh}$$

where the equality is reached for a **GYD**.

Again, it is easy to see that when $\sum_j n_{hj} = r_h$

$$\overline{C}_{hh} \le \bar{\phi}(r_h)$$

and for a **GYD**, $\overline{C}_{hh} = \bar{\phi}(r)$ for all h. Thus, to prove the $\overline{\mathbf{E}}$–optimality of a **GYD**, we need only show that for any competing design

$\bar{\phi}(r_h) \le \bar{\phi}(r)$ for some h.

When the design is not equireplicate, at least one r_h must be less than r and hence it is enough to show that

$$\Delta\bar{g}(x) \ge 0 \text{ for } x < r \tag{5.3.6}$$

where $\Delta\bar{g}(x) = \bar{g}(x+1) - \bar{g}(x)$.

To show that (5.3.6) holds we shall first introduce the following notations.

$$s_1(x) = k(1+2[x/k]),\ s_2(x) = b(1+2[x/b])$$

$$t_1(x) = k^2(1+[x/k])([x/k]),\ t_2(x) = b^2(1+[x/b])([x/b])$$

$$\bar{\alpha} = (1 - \frac{1}{\rho_1})t_1 + (1 - \frac{1}{\rho_2})t_2$$

$$\bar{\beta} = (1 - \frac{1}{\rho_1})(bk-k) + (1 - \frac{1}{\rho_2})(bk-b) - bk(1 - \frac{1}{\rho_1} - \frac{1}{\rho_2})$$

$$\gamma = 1 - \frac{1}{\rho_1} - \frac{1}{\rho_2}.$$

It is easy to see that

$$\Delta\bar{g}(x) = (1 - \frac{1}{\rho_1})(bk-s_1(x)) + (1 - \frac{1}{\rho_2})(bk-s_2(x)) - \gamma(bk-2x-1).$$

From this it can be shown that

$$\Delta\bar{g}(x)>(bk-b-k-2x+1) + (k-1)/\rho_1 + (b-1)/\rho_2.$$

Since each of b and k exceeds one and since $(bk-b-k-2r+1)\geq 0$ when $v\geq 4$ as seen in the fixed effects case, it follows that $\Delta\bar{g}(x)>0$ for $x<r$ and hence a non-regular **GYD** is $\bar{\mathbf{E}}$-optimal.

To examine the $\bar{\mathbf{A}}$- and $\bar{\mathbf{D}}$-optimality of a non-regular **GYD** we note that when $\gamma>0$ the behaviour of $\bar{g}(x+1)-\bar{g}(x)$ is the same as that of $g(x+1)-g(x)$ described in section 3 of Chapter Four. Thus, using the analogy with the fixed effects case we see that to establish $\bar{\mathbf{A}}$-optimality, it is enough to show that

$$\frac{1}{\bar{g}(x+1)} + \frac{1}{\bar{g}(x-1)} \geq \frac{2}{\bar{g}(x)}$$

for $c_0+1\leq x<d_0$ where $[c_0,d_0]$ is the basic interval described in section 3 of Chapter Four. It turns out that this is equivalent to

$$A(x) = 3\gamma^2x^2 + 3\bar{\beta}\gamma x + \bar{\beta}^2 - \bar{\alpha}\gamma - \gamma^2\geq 0$$

for $c_0+1\leq x<d_0$.

Similarly, for $\bar{\mathbf{D}}$-optimality it is enough to show that

$$(\bar{g}(x))^2\geq\bar{g}(x+1)\bar{g}(x-1) \text{ for } c_0+1\leq x<d_0$$

which is equivalent to

$$D(x) = 2\gamma^2x^2 + 2\bar{\beta}\gamma x + \bar{\beta}^2 - 2\bar{\alpha}\gamma - \gamma^2\geq 0$$

for $c_0+1\leq x<d_0$.

Since $\bar{\beta}+2\gamma x>0$ for $c_0+1\leq x<d_0$, $A'(x)>0$ and $D'(x)>0$ for x in that interval. Hence for $\bar{\mathbf{A}}$-optimality ($\bar{\mathbf{D}}$-optimality) it is enough to show that $A(c_0+1)\geq 0$ ($D(c_0+1)\geq 0$).

We note that for the fixed effects case the expressions corresponding to $A(x)$ and $D(x)$ are $\Gamma(x)$ and $\Gamma_0(x)$ defined by (4.3.15) and (4.3.17) respectively. We saw that

$$\Gamma(c_0+1)\geq 0 \text{ for } v\geq 4$$

and

$$\Gamma_0(c_0+1)\geq 0 \text{ for } v\geq 6.$$

Thus, to show that

$$A(c_0+1)\geq 0 \text{ for } v\geq 4$$

and

$$D(c_0+1)\geq 0 \text{ for } v\geq 6$$

it would suffice to show that
(i) $A(c_0+1)\geq D(c_0+1)$ for $v\geq 6$
(ii) $D(c_0+1)-p\Gamma_0(c_0+1)\geq 0$ for $v\geq 6$
and (iii) $A(c_0+1)-q\Gamma(c_0+1)\geq 0$ for $v = 4$
for some $p,q>0$. It may be recalled that for a non-regular **GYD**, $v\geq 4$ and $b_1,b_2\geq 6$.

The above inequalities are established in Mukhopadhyay (1981). However, the computations needed are somewhat lengthy and hence are omitted here.

This completes the proof of the $\bar{\text{A}}$-optimality of any **GYD** and of the $\bar{\text{D}}$-optimality of a **GYD** with $v\geq 6$. These results correspond to the results for the fixed effects model established in Kiefer (1975). It should be noted that even though the methods used in proving these optimality results closely resemble the ones used for the fixed effects case, the derivations are non-trivial and at times quite lengthy.

4. Concluding Remarks

As remarked earlier, the optimality aspects of the designs w.r.t. the mixed effects model have not yet received much attention. It is not known if the specific optimality results for block designs established in Chapter Three hold for the mixed effects model. For row-column designs other than the **GYDs** no optimality results in mixed models are available at this time. Results of sections 2 and 3 seem to indicate that even when the optimality results for the fixed effects model do hold for a mixed model, the proofs may be lengthy and/or difficult.

REFERENCES

Bagchi, S. (formerly Mukhopadhyay, S.) (1987). On the optimality of the MBGDDs under mixed effects model. *Comm. Statist. (A) Theo. Meth., 16*, 3565-3576.

Bhattacharya, C.G. and Shah, K.R. (1984). On the optimality of block designs under a mixed effects model. *Utilitas Mathematica, 26*, 339-345.

Jacroux, M. and Seely, J. (1980). Some sufficient conditions for establishing (M,S)-optimality. *J. Statist. Planning and Inference, 4*, 3-12.

Khatri, C.G. and Shah (1981). Optimality of block designs. Proceedings of the

Indian Statistical Institute Golden Jubilee International Conference on Statistics: Applications and New Directions. Calcutta, December 1981, 326-332.

Kiefer, J. (1975). Construction and optimality of generalized Youden designs. In a *Survey of Statistical Design and Linear Models*. (J.N. Srivastava, ed.). Amsterdam: North-Holland Publishing Co., (1975), 333-353.

Mukhopadhyay, S. (1981). On the optimality of block designs under mixed effects model. *Cal. Statist. Assoc. Bull., 30*, 171-185.

Mukhopadhyay, S. (1984). ψ_f–optimality of the MBGDD of type 1 under the restricted class of binary designs. *Sankhyā (B), 46*, 113-117.

Mukhopadhyay, S. (1987). On the E–optimality of certain asymmetrical designs under mixed effects model. *Metrika, 34*, 95-105.

Rao, C.R. (1947). General methods of analysis for incomplete block designs. *J. Amer. Statist. Assoc., 42*, 541-561.

Roy, J. and Shah, K.R. (1961). Analysis of two-way designs. *Sankhyā (A), 23*, 129-144.

Sinha, B.K. (1980). Optimal Block Designs. *Unpublished Seminar Notes*. Indian Statistical Institute, Calcutta, India.

Yates, F. (1939). The recovery of inter-block information in varietal trials arranged in three dimensional lattices. *Ann. Eugenics, 9*, 136-156

CHAPTER SIX

REPEATED MEASUREMENTS DESIGNS

1. Introduction

In the preceding Chapters, we dealt with optimality aspects of *traditional* block designs and/or row-column designs, assuming fixed/mixed effects models. In many fields of scientific investigations, experiments are to be designed in such a manner that each experimental unit (eu) receives some or all of the treatments, one at a time, over a certain period of time. Such designs have been discussed in the literature under various names, viz., cross-over or change-over designs, time series designs or before-after designs in some special cases. Following Hedayat and Afsarinejad (1975), we will call such designs as *Repeated Measurement Designs* (**RMD**s). In effect, an **RMD** can be viewed as a row-column design with a set of eu's displayed across the columns and a set of periods (of time) displayed across the rows wherein the eu's receive some or all of a given set of treatments, one at a time, over these periods. The *peculiarity* of such an experiment is that any treatment applied to a unit in a certain period *influences* the response of the unit not only in the *current* period but also leaves *residual* effects in the *following* periods. In practice, only the first order residual effect (carry-over effect) i.e., residual effect of any treatment up to just the next period is of importance. For a general review of such designs, including practical applications, reference is made to Hedayat and Afsarinejad (1975). An extreme form of an **RMD** is the one in which *only one* experimental unit is involved in the entire experiment. For such experiments, Finney and Outhwaite (1955, 1956) introduced the notions of serially balanced sequences of types 1 and 2.

The study of optimality aspects of the **RMD**s was initiated by Hedayat and Afsarinejad (1978) and further contributions have been made by Cheng and Wu (1980, 1983), Magda (1980), Dey *et al* (1983) and Kunert (1983, 1984a, 1984b). All these authors considered the problem of characterization and construction of universally optimal designs under fixed effects additive linear models, incorporating effects due to units, periods, and direct and first order residual effects of treatments. In practice, it is not uncommon to face situations where the experimental units involved in the experiment constitute a random sample from a population of a large number of available experimental units. In such a case, a model incorporating random effects due to units has to be sought out. Mukhopadhyay and Saha (1983) took up this study, assuming an additive mixed effects model where period effects and direct and first order residual effects of treatments are regarded as fixed while unit effects are taken as random. They succeeded in extending the optimality results already established in the context of completely fixed effects model to the above mixed effects model. Optimality properties of type 1 and type 2 serially balanced sequences have been investigated in Sinha (1975) and Mukerjee and Sen (1985).

Following these studies, we propose to give a detailed account of hitherto known optimality results concerning the **RMD**s in this Chapter. Of course, as we shall see, the results are available only in restricted set-ups and further research is needed in this area as well. We have organized our presentation as follows. In section 2, we introduce the model(s) along with various definitions and notations peculiar to this experimental set-up. In section 3 we first present various basic facts regarding the combinatorial aspects of *uniform* and/or *strongly balanced* **RMD**s. Then we discuss some algebraic results and, finally, establish universal optimality of strongly balanced uniform **RMDs**.

Section 4 is concerned with some optimality aspects of nearly strongly balanced uniform **RMDs**. In section 5, we discuss results for balanced uniform **RMDs**. Finally, some concluding remarks are given in section 6. Throughout, we will primarily concern ourselves with a fixed-effects additive model. The analogous results for a mixed-effects additive model are mostly stated with reference to the context and these are to be found in Mukhopadhyay and Saha (1983) as also in Saharay (1986).

We will *not* discuss the optimality aspects of serially balanced sequences. We refer the reader to Sinha (1975) and Mukerjee and Sen (1985) for this study. Another related reference is Sen and Sinha (1986).

RMDs with correlated observations have also been studied in the literature. Some useful references are Kunert (1985), Gill and Shukla (1987), Williams (1987) and Street (1988). **RMDs** with autocorrelated errors have been discussed in Azzalini and Giovagnoli (1987) and Mathews (1987). **RMDs** for comparing treatments with a control have been studied in Pigeon and Raghavarao (1987). We will *not* discuss the results of these papers in our monograph.

2. The Linear Model(s), Definitions and Notations

An **RMD** based on v treatments, n experimental units (eu's) and p periods, each unit being given one treatment during each period, is abbreviated as **RMD** (v,n,p). In the *fixed-effects* case, the linear model (1.2.2) still applies. However, this time, $\boldsymbol{\theta}_1$ refers to the $2v\times 1$ vector of direct and residual effects of the treatments under consideration which will be denoted by $\boldsymbol{\tau}$ and $\boldsymbol{\rho}$ respectively. Written explicitly, in the model (1.2.2), the matrix involving $\boldsymbol{\theta}_1$ component now has the representation

$$\mathbf{X}_1 = (\mathbf{T} \vdots \mathbf{F})$$

where $\mathbf{T} = ((t_{(ij)k}))$ and $\mathbf{F} = ((f_{(ij)u}))$ are matrices each of order $np\times v$ indicating respectively the incidences of direct and residual effects of the treatments on the n eu's over the p periods of time. We will denote by $u(ij)$ the treatment applied during the ith period to the jth eu, $1\leq i\leq p$, $1\leq j\leq n$, $1\leq u\leq v$. Clearly, then, $t_{(ij)u(ij)} = 1$, $1\leq i\leq p$, $1\leq j\leq n$, and $f_{(ij)u(i-1,j)} = 1$, $1\leq j\leq n$, $2\leq i\leq p$.

At this stage, we would like to distinguish between

(a) designs with *no* residual effects during the first period and (b) designs with residual effects during the first period.

In case (a), we may set, in the description of $\mathbf{F}$,

$$f_{(1j)u} = 0, \quad 1\leq j\leq n, \quad 1\leq u\leq v. \tag{6.2.1}$$

In case (b), *usually* the residual effects in the first period are derived using those in the last period as the *preperiod* or *conditioning* treatments (Vide Sampford (1957)).

This means that in the description of $\mathbf{F}$, we may set

$$f_{(1j)u(p,j)} = 1, \quad 1 \leq j \leq n \tag{6.2.2}$$

with the understanding that $u(p,j) = u(0,j)$, $1 \leq j \leq n$, "0" representing the *pre-period*.

We will refer to the *model under (a)* as a *non-circular* model. Following Magda (1980) and Kunert (1984b), the *model under (b) with (6.2.2)* will be referred to as a *circular* model.

Again when the eu's involved in the experiment constitute a random sample from a population of a large number of available eu's, we get the set-up of a mixed-effects model in which the period effects and the direct and (first order) residual treatment effects are fixed while the effects of the eu's are random. The underlying mixed-effects model is now similar to the one in (5.3.1). Formally written, this is given by

$$\mathbf{Y} = \mathbf{X}_1\boldsymbol{\theta}_1 + \mathbf{X}_2\boldsymbol{\alpha} + \mathbf{X}_3\boldsymbol{\beta} + \boldsymbol{\epsilon}$$

where, as before, $\boldsymbol{\theta}_1$ represents the $(2v \times 1)$ vector of direct and residual treatment effects. Further, $\boldsymbol{\alpha}$ is the vector of (fixed) period effects while $\boldsymbol{\beta}$ is the vector of (random) unit effects, $\boldsymbol{\epsilon}$ being the vector of observational errors. Here $\mathbf{X}_1$,$\mathbf{X}_2$ and $\mathbf{X}_3$ are matrices of orders $(np \times 2v)$, $(np \times p)$ and $(np \times n)$ respectively. For $\mathbf{X}_1$, we again adopt the representation $\mathbf{X}_1 = (\mathbf{T} \vdots \mathbf{F})$. Then $\mathbf{X}_2'\mathbf{X}_1 = (\mathbf{N}_p' \vdots (\tilde{\mathbf{N}}_p')$ and $\mathbf{X}_3'\mathbf{X}_1 = (\mathbf{N}_u' \vdots \tilde{\mathbf{N}}_u')$ while $\mathbf{X}_2'\mathbf{X}_3 = \mathbf{J}_{p \times n}$. The matrices $\mathbf{N}_p$, $\tilde{\mathbf{N}}_p$, $\mathbf{N}_u$ and $\tilde{\mathbf{N}}_u$ represent respectively the direct treatment vs. period, residual treatment vs. period, direct treatment vs. unit and residual treatment vs. unit incidence matrices.

Regarding $\boldsymbol{\epsilon}$ and $\boldsymbol{\beta}$, we assume that

$$\boldsymbol{\epsilon} \sim N(\mathbf{0},\sigma^2\mathbf{I}_{np}) \text{ and } \boldsymbol{\beta} \sim N(\mathbf{0},\sigma_u^2\mathbf{I}_n)$$

and that $\boldsymbol{\epsilon}$ and $\boldsymbol{\beta}$ are independent.

The (Fisher) information matrix for $\boldsymbol{\theta}_1$ in either model now corresponds to the *joint* information matrix of the direct and residual treatment effects and the optimality studies are to be based on the information matrices separately for the direct and residual treatment effects. For the mixed-effects model, we assume that the variance-ratio σ_u^2/σ^2 is known so that the comparison of competing designs is straightforward.

If we represent $\mathbf{I}(\boldsymbol{\theta}_1)$ formally as $\mathbf{I}(\boldsymbol{\theta}_1) = \begin{pmatrix} \mathbf{I}_{11} & \mathbf{I}_{12} \\ \mathbf{I}_{21} & \mathbf{I}_{22} \end{pmatrix}$, then the two component information matrices are

$$\mathbf{C}(\boldsymbol{\tau}) = \mathbf{I}_{11} - \mathbf{I}_{12}\mathbf{I}_{22}^{-}\mathbf{I}_{21} \text{ (for the direct effects)} \tag{6.2.3}$$

$$C(\rho) = I_{22} - I_{21} I_{11}^{-} I_{12} \text{ (for the residual effects)} \tag{6.2.4}$$

where I_{ii}^{-} refers to a g-inverse of I_{ii} $(i = 1,2)$.

If momentarily we assume the *absence* of the residual effects and work with a fixed-effects model (say), we are in the set-up of a row-column design with the direct effects of the treatments as the only treatment effects. Hence, in such a situation, the information matrix for the treatment effects will have a representation of the form (1.2.5). This will certainly then correspond to the component I_{11} shown above. A similar consideration applies to I_{22} as well. This sort of argument leads to the expressions for I_{11} and I_{22} directly. Of course, there is no such convenient argument for deriving the expression for I_{12}. One has to refer to the entire expression for $I(\theta_1)$ derivable from (1.2.3). The details of such calculations in the fixed-effects model are to be found in Cheng and Wu (1980) as also in Dey *et al* (1983). The analogous expressions in the mixed-effects model are given in Mukhopadhyay and Saha (1983). We will reproduce all the expressions explicitly later in this section.

Our study will bring out optimality properties of some classes of **RMDs** having nice combinatorial features as in the case with the **BBDs** and the **GYDs** in traditional set-up of block designs and row-column designs. For example, among other things, we will establish universal optimality of what are called *strongly balanced uniform* **RMDs**. To facilitate the presentation, we now define various terms used in this context. The set-up is one of an **RMD** with parameters v, n and p.

Definitions

(1) A design is called *uniform on periods* if each treatment occurs the same number of times, say λ_1 times in each period. A necessary condition for this to hold is $n = \lambda_1 v$.

(2) A design is called *uniform on units* if each treatment is assigned the same number of times, say, λ_2 times to each unit. This can occur only if $p = \lambda_2 v$.

(3) A design is called *uniform* if it is uniform on both periods and units. A uniform **RMD** $(v,\lambda_1 v,\lambda_2 v)$ is denoted by the symbol **URMD** $(v,\lambda_1 v,\lambda_2 v)$.

(4) A design underlying a non-circular (circular) model is called *balanced* if the collection of ordered pairs $(u(i,j),u(i-1,j))$ for $j = 1,2,\ldots,n$ and $i = 2,3,\ldots,p$ in a non-circular model (respectively for $i = 1,2,\ldots,p$ in a circular model) contain each *ordered* distinct pair of treatments equal number of times, say, λ times. For a *non-circular model* a necessary condition for this to hold is $n(p-1) = \lambda v(v-1)$. For a *circular model* the corresponding condition is $np = \lambda v(v-1)$.

(5) A design is called *strongly balanced* if the ordered pairs $(u(i,j),u(i-1,j))$ as given in definition (4) contain each ordered pair (including pairs of identical treatments) equal number of times, say, λ times.

For a non-circular model, a necessary condition for this to hold is $n(p-1) = \lambda v^2$. For a circular model, the corresponding condition is $np = \lambda v^2$.

Some further definitions will be incorporated later. We now fix various notations to be used in the present context. Most of the notations are analogues of those needed in the study of a row-column design. Here we need extra considerations as both direct and residual effects of treatments are to be handled simultaneously.

Notations

$n_{hj} =$ number of appearances of treatment h on unit j in the periods 1 to p,

$\tilde{n}_{hj} =$ number of appearances of treatment h on unit j in the periods 0 to $p-1$,

$l_{hi} =$ number of appearances of treatment h in period i over the units 1 to n

$\tilde{l}_{hi} =$ number of appearances of treatment h in period $i-1$ over the units 1 to n,

$m_{hh'} =$ number of appearances of treatment h preceded by treatment h' on the same unit and summed over all units,

$r_h =$ number of appearances of treatment h in the periods 1 to p over the units 1 to n,

$\tilde{r}_h =$ number of appearances of treatment h in the periods 0 to $p-1$ over the units 1 to n,

$s_h =$ number of appearances of treatment h in the periods 2 to p over the units 1 to n,

where $1 \le j \le n$, $1 \le i \le p$, $1 \le h$, $h' \le v$.

Then the following relations are immediate.

$$r_h = \sum_{j=1}^{n} n_{hj} = \sum_{i=1}^{p} l_{hi}$$

$$\tilde{r}_h = \sum_{j=1}^{n} \tilde{n}_{hj} = \sum_{i=1}^{p} \tilde{l}_{hi} = \sum_{h'=1}^{v} m_{h'h} \tag{6.2.5}$$

$$s_h = \sum_{i=2}^{p} l_{hi} = \sum_{h'=1}^{v} m_{hh'} \quad \text{(for a non-circular model)} \tag{6.2.6}$$

$$\sum_{h=1}^{v} n_{hj} = p, \ \sum_{h=1}^{v} l_{hi} = n, \ \sum_{h=1}^{v} r_h = np$$

$$\sum_{h=1}^{v} \tilde{n}_{hj} = p-1, \ \sum_{h=1}^{v} \tilde{r}_h = n(p-1) \ \text{(for a non-circular model).}$$

Let $\mathbf{r} = (r_1, r_2, \ldots, r_v)'$ and $\tilde{\mathbf{r}} = (\tilde{r}_1, \tilde{r}_2, \ldots, \tilde{r}_v)'$. Define now the following matrices:

$$\mathbf{r}^{\delta} = Diag(r_1, \ldots, r_v), \ \tilde{\mathbf{r}}^{\delta} = Diag(\tilde{r}_1, \ldots, \tilde{r}_v)$$

$$\mathbf{M} = ((m_{hh'})), \ 1 \le h, h' \le v$$

$$\mathbf{N}_p = ((l_{hi})), \ 1 \le h \le v, \ 1 \le i \le p$$

$$\tilde{\mathbf{N}}_p = ((\tilde{l}_{hi})), \ 1 \le h \le v, \ 1 \le i \le p$$

$$\mathbf{N}_u = ((n_{hj})), \ 1 \le h \le v, \ 1 \le j \le n$$

$$\tilde{\mathbf{N}}_u = ((\tilde{n}_{hj})), \ 1 \le h \le v, \ 1 \le j \le n.$$

Also **1** will represent a column vector (of appropriate order) of 1's and $\mathbf{J}_{p \times q}$ will represent a matrix of order $p \times q$ of all 1's. When $p = q$, we write it as $\mathbf{J}_p$. As stated earlier, it may be noted that $\mathbf{N}_p(\tilde{\mathbf{N}}_p)$ and $\mathbf{N}_u(\tilde{\mathbf{N}}_u)$ denote respectively the direct (residual) treatment vs. period and direct (residual) treatment vs. unit incidence matrices when the **RMD** is regarded as a row-column design. On the other hand, recalling that $\mathbf{T}$ and $\mathbf{F}$ are matrices showing the incidences of direct and residual effects of the treatments on the eu's, we easily verify that $\mathbf{T}'\mathbf{F} = \mathbf{M}$. The element $m_{hh'}$ of $\mathbf{M}$ stands for the number of times the direct effect of hth treatment (τ_h) and the residual effect of h'th treatment $(\rho_{h'})$ occur simultaneously and explicitly in the linear model under consideration (for both fixed- and mixed-effects models) in the form $\tau_h + \rho_{h'}$.

We are now in a position to write down the explicit expressions for $\mathbf{I}_{11}, \mathbf{I}_{22}$ and $\mathbf{I}_{12}$ under both fixed- and mixed-effects models. To differentiate between the two types of models, we will use the notations $\mathbf{I}_{ij}(F)$ and $\mathbf{I}_{ij}(M)$ where 'F' and 'M' refer to fixed- and mixed-effects models respectively. Our previous analysis as to the interpretation of $\mathbf{I}_{11}$ and $\mathbf{I}_{22}$ yield readily the following expressions. Recall (1.2.5) in this context.

$$\mathbf{I}_{11}(F) = \mathbf{r}^{\delta} - n^{-1}\mathbf{N}_p\mathbf{N}_p' - p^{-1}\mathbf{N}_u\mathbf{N}_u' + n^{-1}p^{-1}\mathbf{r}\mathbf{r}' \tag{6.2.7}$$

$$\mathbf{I}_{22}(F) = \tilde{\mathbf{r}}^{\delta} - n^{-1}\tilde{\mathbf{N}}_p\tilde{\mathbf{N}}_p' - p^{-1}\tilde{\mathbf{N}}_u\tilde{\mathbf{N}}_u' + n^{-1}p^{-1}\tilde{\mathbf{r}}\tilde{\mathbf{r}}'. \tag{6.2.8}$$

The expression for $\mathbf{I}_{12}(F)$ follows from a direct computation of $\mathbf{I}(\boldsymbol{\theta}_1)$ using the general expression (1.2.3). This is done in Cheng and Wu (1980). See also Dey *et al* (1983). We give the expression below.

$$\mathbf{I}_{12}(F) = \mathbf{M}-n^{-1}\mathbf{N}_p\tilde{\mathbf{N}}_p'-p^{-1}\mathbf{N}_u\tilde{\mathbf{N}}_u'+n^{-1}p^{-1}\mathbf{r}\tilde{\mathbf{r}}'. \tag{6.2.9}$$

These expressions for $\mathbf{I}_{11}$, $\mathbf{I}_{22}$ and $\mathbf{I}_{12}$ refer to the usual fixed-effects model. In the mixed-effects model, the analogous expressions are

$$\mathbf{I}_{11}(M) = \mathbf{r}^{\delta}-n^{-1}\mathbf{N}_p\mathbf{N}_p'-\frac{w-\tilde{w}}{w}\{p^{-1}\mathbf{N}_u\mathbf{N}_u'-n^{-1}p^{-1}\mathbf{r}\mathbf{r}'\} \tag{6.2.7$'$}$$

$$\mathbf{I}_{22}(M) = \tilde{\mathbf{r}}^{\delta}-n^{-1}\tilde{\mathbf{N}}_p\tilde{\mathbf{N}}_p'-\frac{w-\tilde{w}}{w}\{p^{-1}\tilde{\mathbf{N}}_u\tilde{\mathbf{N}}_u'-n^{-1}\tilde{\mathbf{r}}\tilde{\mathbf{r}}'\} \tag{6.2.8$'$}$$

$$\mathbf{I}_{12}(M) = \mathbf{M}-n^{-1}\mathbf{N}_p\tilde{\mathbf{N}}_p'-\frac{w-\tilde{w}}{w}\{p^{-1}\mathbf{N}_u\tilde{\mathbf{N}}_u'-n^{-1}p^{-1}\mathbf{r}\tilde{\mathbf{r}}'\} \tag{6.2.9$'$}$$

where $w = \sigma^{-2}$ and $\tilde{w} = (\sigma^2+p\sigma_u^2)^{-1}$ so that $w>\tilde{w}$.

We refer to Mukhopadhyay and Saha (1983) for this and other derivations in mixed-effects model.

The information matrices $\mathbf{C}(\tau)$ and $\mathbf{C}(\rho)$ for the direct and residual treatment effects can now be written down using (6.2.3) and (6.2.4) respectively.

The optimality results for the **RMDs** so far discussed in the literature deal only with universal optimality in the sense of Kiefer and, accordingly, rest on the verification of the (sufficient) conditions stated in Proposition 1 of Kiefer (1975). However, as we have noted explicitly, the same conditions enable us to establish extended universal optimality as well. Therefore, the optimality results to be presented below apply to universal optimality in an extended sense. The verification, as we know, is in terms of trace maximization and complete symmetry.

3. Universal Optimality of Strongly Balanced Uniform RMDs

In this section, we first present certain basic facts regarding the nature of some classes of the **RMDs**. These will facilitate presenting the results regarding optimality of such **RMDs** in the entire class or in an appropriate subclass of the **RMDs** for given v,n and p. At this stage, we may recall the definition of a Generalized Youden Design (**GYD**) as given in Chapter Four. A regular **GYD** is a **GYD** with $v|n$ and/or $v|p$. Kunert (1983) calls a regular **GYD** a Generalized Latin Square (**GLS**) when $v|n$ *and* $v|p$. An **RMD** with parameters v,n and p has already been abbreviated as **RMD** (v,n,p).

(i) For a balanced **RMD**, $\mathbf{M} \propto \mathbf{J}_v-\mathbf{I}_v$ and for a strongly balanced **RMD**, $\mathbf{M} \propto \mathbf{J}_v$. This is true for *both* circular and non-circular models.

(ii) For a uniform **RMD** $(v,\lambda_1 v,\lambda_2 v)$, $\mathbf{r} = p\lambda_1\mathbf{1}$, $\tilde{\mathbf{r}} = \lambda_1(p-1)\mathbf{1}$, $\mathbf{N}_p = \lambda_1\mathbf{J}$, $\mathbf{N}_u = \lambda_2\mathbf{J}$, $\tilde{\mathbf{N}}_p\tilde{\mathbf{N}}_p' = \lambda_1^2(p-1)\mathbf{J}_v$, $\tilde{\mathbf{N}}_u\tilde{\mathbf{N}}_u' = (n\lambda_2^2-2\lambda_1\lambda_2)\mathbf{J}_v+\lambda_1\mathbf{I}_v$, $\mathbf{N}_p\tilde{\mathbf{N}}_p' = \lambda_1^2(p-1)\mathbf{J}_v$ and $\mathbf{N}_u\tilde{\mathbf{N}}_u' = \lambda_1\lambda_2(p-1)\mathbf{J}_v$ for a non-circular model. For a circular model, the corresponding expressions are deduced from $\mathbf{r} = \tilde{\mathbf{r}} = \lambda_1 p\mathbf{1}$, $\mathbf{N}_p = \tilde{\mathbf{N}}_p = \lambda_1\mathbf{J}$, $\mathbf{N}_u = \tilde{\mathbf{N}}_u = \lambda_2\mathbf{J}$.

(iii) For an **RMD** $(v,\lambda_1 v,\lambda_2 v+1)$ which is uniform on the periods and also uniform on the units in the first $(p-1)$ periods, $\mathbf{r} = p\lambda_1\mathbf{1}$, $\tilde{\mathbf{r}} = \lambda_1(p-1)\mathbf{1}$, $\mathbf{N}_p = \lambda_1\mathbf{J}$, $\tilde{\mathbf{N}}_u = \lambda_2\mathbf{J}$, $\tilde{\mathbf{N}}_p\tilde{\mathbf{N}}_p' = \lambda_1^2(p-1)\mathbf{J}_v$, $\mathbf{N}_u\mathbf{N}_u' = (n\lambda_2^2+2\lambda_1\lambda_2)\mathbf{J}_v+\lambda_1\mathbf{I}_v$, $\mathbf{N}_u\tilde{\mathbf{N}}_u' = (n\lambda_2^2+\lambda_1\lambda_2)\mathbf{J}_v$ and $\mathbf{N}_p\tilde{\mathbf{N}}_p' = \lambda_1^2(p-1)\mathbf{J}_v$ for a non-circular model.

(iv) Regarded as a row-column design for direct treatment effects, every uniform **RMD** $(v,\lambda_1 v,\lambda_2 v)$ is a **GLS**. Consequently, $\mathbf{I}_{11}(F)$ in (6.2.7) is completely symmetric. Further, among *all* **RMDs**, such uniform **RMDs** maximize $tr\{\mathbf{I}_{11}(F)\}$. (For this last statement of fact, recall the first step towards establishing universal optimality of the regular **GYDs** as deduced in section 2, Chapter Four). The same fact also applies to $\mathbf{I}_{11}(M)$ in (6.2.7)$'$. This is true for both circular and non-circular models.

(v) With reference to a non-circular (circular) model, regarded as a row-column design for residual treatment effects, every uniform **RMD** $(v,\lambda_1 v,\lambda_2 v)$ is a **GYD** (**GLS**). Consequently, $\mathbf{I}_{22}(F)$ in (6.2.8) is completely symmetric. Further, among *all* **RMDs**, such uniform **RMDs** maximize $tr\{\mathbf{I}_{22}(F)\}$. The same fact also applies to $\mathbf{I}_{22}(M)$.

(vi) For a strongly balanced **RMD** $(v,\lambda_1 v,\lambda_2 v)$, $\mathbf{M} = \lambda\mathbf{J}$ where $\lambda = n(p-1)/v^2 = \lambda_1(p-1)/v$ for a non-circular model and $\lambda = np/v^2 = \lambda_1\lambda_2$ for a circular model. Hence, for a strongly balanced uniform **RMD** $(v,\lambda_1 v,\lambda_2 v)$, the expression for $\mathbf{I}_{12}(F)$ in (6.2.9) simplifies to

$$\mathbf{I}_{12}(F) = \begin{cases} g_n\mathbf{J}_v & \text{under a non-circular model} \\ g_c\mathbf{J}_v & \text{under a circular model} \end{cases}$$

where

$$g_n = \{\lambda_1(p-1)/v\} - n^{-1}\lambda_1^2(p-1) - p^{-1}\lambda_1\lambda_2(p-1) + n^{-1}p^{-1}\lambda_1^2 p(p-1)$$

and

$$g_c = \lambda_1\lambda_2 - n^{-1}\lambda_1^2 p - p^{-1}\lambda_2^2 n + \lambda_1\lambda_2.$$

Now it is readily seen that each of the constants g_n and g_c simplifies to zero so that $\mathbf{I}_{12}(F)$ reduces to a null matrix under both non-circular and circular models. The same phenomenon holds also for $\mathbf{I}_{12}(M)$ in (6.2.9)$'$. In other words, $\mathbf{I}_{12}(M) = 0$ for a strongly balanced uniform **RMD** $(v,\lambda_1,v,\lambda_2,v)$ under both non-circular and circular models.

(vii) For a strongly balanced **RMD** $(v,\lambda_1 v,\lambda_2 v+1)$ which is uniform on the periods and also uniform on the units in the first $(p-1)$ periods, we have $\mathbf{M} = \lambda\mathbf{J}$ where $\lambda = n(p-1)/v^2 = \lambda_1\lambda_2$ for a non-circular model. Further, referring to (iii) above, we have the following expression for $\mathbf{I}_{12}(F)$ in (6.2.9) for a non-circular model.

$$\mathbf{I}_{12}(F) = \lambda_1\lambda_2\mathbf{J}_v - n^{-1}\lambda_1^2\lambda_2 v\mathbf{J}_v - p^{-1}\lambda_2(n\lambda_2+\lambda_1)\mathbf{J}_v + n^{-1}p^{-1}p\lambda_1^2(p-1)\mathbf{J}_v.$$

Once more, the scalar multiplier of $\mathbf{J}_v$ simplifies to zero! Thus, $\mathbf{I}_{12}(F) = \mathbf{0}$ in such a case. A similar analysis suggests that $\mathbf{I}_{12}(M) = \mathbf{0}$ as well. Thus, for a non-circular model, both $\mathbf{I}_{12}(F)$ and $\mathbf{I}_{12}(M)$ reduce to null matrices for a strongly balanced **RMD** $(v,\lambda_1 v,\lambda_2 v+1)$ which is uniform on the periods and also uniform on the units in the first $(p-1)$ periods.

We are now in a position to present the following universal optimality results using the above facts.

Theorem 6.3.1. For given (v,n,p), a strongly balanced uniform **RMD** (v,n,p), whenever it exists, is universally optimal for the estimation of direct as well as residual effects among *all* relevant competing designs under both circular/non-circular and fixed-effects/mixed-effects models.

Proof. The proof is based on the facts stated in (iv), (v) and (vi) above. To see this, note from (6.2.3) and (6.2.4) that for any competing design $\mathbf{C}(\tau\,|.) \le \mathbf{I}_{11}(.)$ and $\mathbf{C}(\rho\,|.) \le \mathbf{I}_{22}(.)$ with equality for a strongly balanced uniform **RMD** (v,n,p) which also maximizes $tr\{\mathbf{I}_{11}(.)\}$ and $tr\{\mathbf{I}_{22}(.)\}$. This is true for both circular/non-circular and fixed-/mixed-effects models. Hence the claim.

Theorem 6.3.2. For given (v,n,p) with $v\,|\,n$ and $p = 1(mod\ v)$, a strongly balanced **RMD** (v,n,p) which is uniform on the periods and also uniform on the units in the first $p-1$ $(\ge v)$ periods, whenever it exists, is universally optimal for the estimation of direct as well as residual effects among *all* relevant competing designs under non-circular fixed-/mixed-effects models.

Proof. The proof is again based on the facts stated in (vii). We omit the steps already outlined in the proof of Theorem 6.3.1.

Remark 6.3.1. The constructional aspects of strongly balanced uniform **RMDs** have been discussed, among others, by Hedayat (1981), Constantine and Hedayat (1982), Afsarinejad (1983), Kunert (1983), Sen and Mukerjee (1987) and Roy (1988). The last three articles deal with optimality and constructional aspects of **RMDs** in the presence of possible interaction between direct and residual effects of treatments. As to the constructional aspects of optimal designs of the type suggested under Theorem 6.3.2, Cheng and Wu (1980) provide a quick example for the case of $p = v+1$. Suppose there is a balanced uniform **RMD** $(v,\lambda_1 v,v)$. Then by repeating the arrangement of the treatments in the last period of the above **RMD** $(v,\lambda_1 v,v)$ to the next period, one can generate an **RMD** $(v,\lambda_1 v,v+1)$ satisfying the conditions of Theorem 6.3.2. Moreover, from the definition of a strongly balanced uniform **RMD** (v,n,p), it is clear that the existence of such a design necessitates $v^2|\,n$. Suppose in the general case of $p>v+1$, we start with such a design and add an extra period such that the treatment pairs in the pth and $(p+1)$th periods cover all pairs of the type (h,h') equally often, $1\le h,\ h'\le v$. We can easily verify that the resulting design is again strongly balanced and satisfies the conditions of Theorem 6.3.2. This method of construction is also suggested in Cheng and Wu (1980).

Remark 6.3.2. Referring to (6.2.7) as also to (6.2.7)$'$, we observe that for any *equireplicate* competing design for given v, n and p, $\mathbf{I}_{11}(.) \le r\mathbf{I}_v - n^{-1}\mathbf{N}_p\mathbf{N}_p'$ with *equality* for a strongly balanced uniform **RMD** (v,n,p) and, for such a design $\mathbf{I}_{11}(.)$ is precisely given by $r(\mathbf{I}_v - \mathbf{J}_v/v)$ where $r = np/v$. The above analysis holds for both circular and non-circular models. As to the validity of a similar analysis for $\mathbf{I}_{22}(.)$ in (6.2.8) and (6.2.8)′, we need further to assume that the competing designs are *equireplicate* in the last period (and, hence, consequently, in the first $(p-1)$ periods) so far as the non-circular model is concerned. This assumption now enables us to claim stronger and striking optimality results for strongly balanced uniform **RMDs** as discussed in Cheng and Wu (1980), Magda (1980), Mukhopadhyaya and Saha (1983) and Saharay (1986). As a matter of fact, this is the kind of result which induces total dominance of the **C**-matrix of an optimal design over that of *any* of its competitors. Such results are very rarely found in the literature.

We enunciate the following.

Theorem 6.3.3. For given (v,n,p), a strongly balanced uniform **RMD** (v,n,p), whenever it exists, minimizes the variance of the best linear unbiased estimator of *any* contrast involving the *direct* effects among *all* relevant competing *equireplicate* designs under both circular/non-circular and fixed-/mixed-effects models.

Proof. By our analysis in Remark 6.3.2, for any competing design, $\mathbf{C}(\tau\,|.) \leq \mathbf{I}_{11}(.) \leq r\mathbf{I}_v - n^{-1}\mathbf{N}_p\mathbf{N}_p'$ so that for any given $v\times 1$ vector $\mathbf{x}$ s.t. $\mathbf{x}'\mathbf{1} = 0$, $\mathbf{x}'\mathbf{C}(\tau\,|.)\mathbf{x} \leq r\mathbf{x}'\mathbf{x}$. On the other hand, for a strongly balanced uniform **RMD** (v,n,p), $\mathbf{C}(\tau\,|.) = \mathbf{I}_{11}(.) = r(\mathbf{I}_v - \mathbf{J}_v/v)$ as $\mathbf{I}_{12}(.) = \mathbf{0}$ and, hence, for such a design, $\mathbf{x}'\mathbf{C}(\tau\,|.)\mathbf{x} = r\mathbf{x}'\mathbf{x}$ for any $\mathbf{x}$ s.t. $\mathbf{x}'\mathbf{1} = 0$. Writing $\mathbf{C}_1$ for the $\mathbf{C}(\tau\,|.)$-matrix of a competing design and $\mathbf{C}_2$ for that of a strongly balanced uniform **RMD** (v,n,p), we have thus deduced that $\mathbf{x}'\mathbf{C}_1\mathbf{x} \leq \mathbf{x}'\mathbf{C}_2\mathbf{x}$ and, hence, that $\mathbf{x}'\mathbf{C}_1^{+}x \geq \mathbf{x}'\mathbf{C}_2^{+}\mathbf{x}$ for any $\mathbf{x}$ satisfying $\mathbf{x}'\mathbf{1} = 0$. Since the variance of *any* estimable direct effects treatment contrast is proportional to $\mathbf{x}'\mathbf{C}^{+}(\tau\,|.)\mathbf{x}$ for *some* $\mathbf{x}$ satisfying $\mathbf{x}'\mathbf{1} = 0$, the result follows.

Theorem 6.3.4. For given (v,n,p), a strongly balanced uniform **RMD** (v,n,p), whenever it exists, minimizes the variance of the best linear unbiased estimator of *any* contrast involving the *residual* effects among *all* relevant equireplicate competing designs under circular fixed-/mixed-effects models. The same is true also of non-circular fixed-/mixed-effects models provided the competing (equireplicate) designs are also equireplicate in the last period.

Proof. The proof follows along the same line of arguments as in the previous Theorem. We omit the details.

Remark 6.3.3. So far we have worked with two types of models viz., non-circular and circular. Kunert (1983) considered another model in which there is a pre-period but the arrangement of treatments in the pre-period is completely arbitrary. For such a model, he deduced universal optimality of some classes of uniform **RMDs** for the estimation of direct effects under a fixed-effects model. The result is indeed true under a mixed-effects model as well. The precise statement is given below.

Theorem 6.3.5. For given (v,n,p), suppose there exists a uniform **RMD** (v,n,p) for which $m_{hh'} = \tilde{r}_{h'}/v$, $1 \leq h,\ h' \leq v$. Then such a design is universally optimal for the estimation of direct effects under both fixed- and mixed-effects models in a design set-up where there is a pre-period, the arrangement of the treatments in the pre-period being arbitrary.

Proof. It is enough to observe that for the uniform **RMD** (v,n,p) under consideration, $\mathbf{M} = \mathbf{1}\tilde{\mathbf{r}}'/v$ (from the given values of $m_{hh'}$'s), $\mathbf{N}_p\tilde{\mathbf{N}}_p' = n\mathbf{1}\tilde{\mathbf{r}}'/v$, $\mathbf{N}_u\tilde{\mathbf{N}}_u' = p\mathbf{1}\tilde{\mathbf{r}}'/v$ (Vide (ii) for the details). Hence, referring to (6.2.9) and (6.2.9)′, we deduce that

$\mathbf{I}_{12}(F) = \mathbf{I}_{12}(M) = \mathbf{0}$. Next, we observe that $\mathbf{I}_{11}(F)$ and $\mathbf{I}_{11}(M)$ in (6.2.7) and (6.2.7)′ respectively, are independent of the arrangement of the treatments in the pre-period. It is now evident that the uniform **RMD** is universally optimal (since we are in a set-up where $v|n$ and $v|p$).

Remark 6.3.4. If $\tilde{r}_h$'s are all equal, the condition of Theorem 6.3.5 amounts to strong balance for the optimal uniform **RMD** (v,n,p). Thus, the optimality result for a circular model follows as a special case. However, for a design with arbitrary pre-period arrangement, this result is quite general in nature.

An optimalty result slightly more general than the above for a design set-up with an arbitrary pre-period arrangement is also given by Kunert (1983). It essentially follows from the expressions (6.2.9) and (6.2.9)′ for $\mathbf{I}_{12}(.)$. This time we demand that for the optimal design $\mathbf{M} = n^{-1}\mathbf{N}_p\tilde{\mathbf{N}}_p'$ and $\mathbf{N}_u\tilde{\mathbf{N}}_u' = n^{-1}\mathbf{r}\tilde{\mathbf{r}}'$ so that $\mathbf{I}_{12}(.)$ again reduces to a null matrix. It is easy to see that the second requirement is satisfied by any design which is uniform on units. We simply state the result below (covering also the mixed-effects model).

Theorem 6.3.6. For given (v,n,p) with $v \nmid n$ and $v|p$, suppose there exists an **RMD** (v,n,p) which is uniform on units and for which $\mathbf{M} = n^{-1}\mathbf{N}_p\tilde{\mathbf{N}}_p'$. Then such a design is universally optimal for the estimation of direct effects under both fixed- and mixed-effects models in a design set-up where there is a pre-period, the arrangements of the treatments in the pre-period being arbitrary.

Remark 6.3.5. In the statement of the above Theorem, one could replace the condition $\mathbf{M} = n^{-1}\mathbf{N}_p\tilde{\mathbf{N}}_p'$ by $\mathbf{M} = p^{-1}\mathbf{N}_u\tilde{\mathbf{N}}_u'$. But this would give the result only under a fixed-effects model. The difficulty to extend the result to a mixed-effects model is clear from the expression for $\mathbf{I}_{12}(M)$ in (6.2.9)′.

Kunert (1983) has also given some sporadic examples of such optimal designs.

4. Universal Optimality of Nearly Strongly Balanced Uniform RMDs

So far we have primarily considered optimality of strongly balanced uniform **RMD**s which exist only if $v^2|n$ and $p \geq 2v$ whenever the model is a non-circular one. We now examine the situations where $p \geq 2v$ and $v|n$ but $v^2 \nmid n$. We take $n = Av^2+Bv$, $A \geq 1$, $1 \leq B \leq v-1$ and proceed to present results on optimality of what are termed *nearly strongly balanced* uniform **RMD**s, under a non-circular model. We start with the following:

Definition

For given (v,n,p), an **RMD** (v,n,p) is called nearly strongly balanced if $\mathbf{MM}'$ is completely symmetric and $m_{hh'}$'s assume values $[n(p-1)/v^2]$ or $[n(p-1)/v^2]+1$, $1 \leq h$, $h' \leq v$.

We now study some properties of uniform **RMD**s in relation to their competitors in a suitable subclass of the **RMD**s. Consider the subclass of *all* **RMD**s which are *uniform on units and on the last period*. Then, referring to (6.2.7)–(6.2.9) and (6.2.7)′–(6.2.9)′, we deduce that for any such competing design

$$\left.\begin{aligned}
\mathbf{I}_{11}(F) &= \mathbf{I}_{11}(M) = n\lambda_2\mathbf{I}_v - n^{-1}\mathbf{N}_p\mathbf{N}_p' \\
\mathbf{I}_{12}(F) &= \mathbf{I}_{12}(M) = \mathbf{M} - n^{-1}\mathbf{N}_p\tilde{\mathbf{N}}_p' \\
\mathbf{I}_{22}(F) &= \{\lambda_1(p-1) - \frac{\lambda_1}{p}\}\mathbf{I}_v + \frac{\lambda_1\mathbf{J}_v}{pv} - n^{-1}\tilde{\mathbf{N}}_p\tilde{\mathbf{N}}_p' \\
\mathbf{I}_{22}(M) &= \{\lambda_1(p-1) - \frac{\lambda_1(w-\tilde{w})}{pw}\}\mathbf{I}_v + \frac{(w-\tilde{w})\lambda_1\mathbf{J}_v}{wpv} - n^{-1}\tilde{\mathbf{N}}_p\tilde{\mathbf{N}}_p'
\end{aligned}\right\} \quad (6.4.1)$$

when w and $\tilde{w}$ are functions of the variance components σ^2 and σ_u^2 defined earlier. In deducing these results, we have used the fact that $\mathbf{N}_u = \lambda_2\mathbf{J}$, $\mathbf{r} = n\lambda_2\mathbf{1}$, $\tilde{\mathbf{r}} = \lambda_1(p-1)\mathbf{1}$, $\tilde{\mathbf{N}}_u\tilde{\mathbf{N}}_u' = \lambda_1\lambda_2(p-2)\mathbf{J}_v + \lambda_1\mathbf{I}_v$. We will now re-write $\mathbf{I}_{11}(.)$ and $\mathbf{I}_{22}(.)$ in convenient forms. Note that $\mathbf{N}_p\mathbf{1} = p\lambda_1\mathbf{1}$ for any competing design. Thus $\mathbf{I}_{11}(F) = n\lambda_2(\mathbf{I}_v - \mathbf{J}_v/v) - n^{-1}\mathbf{N}_p(\mathbf{I}_p - \mathbf{J}_p/p)\mathbf{N}_p' = \mathbf{I}_{11}(M)$. Again, for a non-circular model, we may write $\tilde{\mathbf{N}}_p = (\mathbf{0}|\tilde{\mathbf{N}}_p^*)$ so that $\tilde{\mathbf{N}}_p\tilde{\mathbf{N}}_p' = \tilde{\mathbf{N}}_p^*\tilde{\mathbf{N}}_p^{*'} = \tilde{\mathbf{N}}_p^*(\mathbf{I}_{p-1} - \frac{\mathbf{J}_{p-1}}{p-1})\tilde{\mathbf{N}}_p^{*'} + \frac{\tilde{\mathbf{r}}\tilde{\mathbf{r}}'}{p-1}$. Using this representation, we may rewrite $\mathbf{I}_{22}(.)$ as

$$\mathbf{I}_{22}(F) = \{\lambda_1(p-1) - \frac{\lambda_1}{p}\}(\mathbf{I}_v - \mathbf{J}_v/v) - n^{-1}\tilde{\mathbf{N}}_p^*(\mathbf{I}_{p-1} - \frac{\mathbf{J}_{p-1}}{p-1})\tilde{\mathbf{N}}_p^{*'}$$

$$\mathbf{I}_{22}(M) = \{\lambda_1(p-1) - \frac{\lambda_1(w-\tilde{w})}{pw}\}(\mathbf{I}_v - \frac{\mathbf{J}_v}{v}) - n^{-1}\tilde{\mathbf{N}}_p^*(\mathbf{I}_{p-1} - \mathbf{J}_{p-1}/(p-1))\tilde{\mathbf{N}}_p^{*'}$$

Clearly, then, for any **RMD** which is uniform on units and on the last period,

$$\left.\begin{aligned}
\mathbf{I}_{11}(F) &= \mathbf{I}_{11}(M) \le n\lambda_2(\mathbf{I}_v - \mathbf{J}_v/v) \\
\mathbf{I}_{22}(F) &\le \{\lambda_1(p-1) - \frac{\lambda_1}{p}\}(\mathbf{I}_v - \mathbf{J}_v/v) \\
\mathbf{I}_{22}(M) &\le \{\lambda_1(p-1) - \frac{\lambda_1(w-\tilde{w})}{pw}\}(\mathbf{I}_v - \mathbf{J}_v/v)
\end{aligned}\right\} \quad (6.4.2)$$

Further, in all of the above, equality holds for *any uniform* **RMD**.

Various optimality results now follow and the proofs get more and more involved as we go for wider subclasses of competing designs. The first result is quite straightforward.

Theorem 6.4.1. For given (v,n,p), suppose there exists a nearly strongly balanced uniform **RMD** (v,n,p). Then it is universally optimal under both fixed- and mixed-effects models for the estimation of direct as well as residual effects among all competing uniform **RMD**s.

Proof. For *any* competing uniform **RMD**, equality holds everywhere in (6.4.2).

Thus, referring to (6.2.3) and (6.2.4) and using the fact that for an nnd matrix $\mathbf{A} = a(\mathbf{I}_v - \mathbf{J}_v/v)$, $\mathbf{A}^+ = a^{-1}\mathbf{I}_v$, it is seen that a uniform **RMD** will maximize the trace of $\mathbf{C}(\tau|.)$ or $\mathbf{C}(\rho|.)$ iff it minimizes $tr\{\mathbf{I}_{12}^2(.)\}$. However, $tr(\mathbf{I}_{12}^2(.)$

$= tr(\{\mathbf{M}-\lambda_1^2(p-1)\mathbf{J}_v\}^2) = \sum_h\sum_{h'}\{m_{hh'}-\lambda_1^2(p-1)\}^2$. Recall that (Vide (6.2.5)) $\sum_h m_{hh'} = \bar{r}_{h'}$ $= \lambda_1(p-1) = \frac{n(p-1)}{v}$ for $1 \le h' \le v$. Hence we can write $\sum_h\sum_{h'}\{m_{hh'}-\lambda_1^2(p-1)\}^2$ $= \sum_{h'}[\sum_h (m_{hh'}-\lambda_1^2(p-1))^2] = \sum_{h'}[\sum_h (m_{hh'} - \frac{n(p-1)}{v^2})^2 + v(\frac{n(p-1)}{v^2} - \lambda_1^2(p-1))^2]$ $= v^2[\frac{n(p-1)}{v^2}-\lambda_1^2(p-1)]^2 + \sum_h\sum_{h'}(m_{hh'} - \frac{n(p-1)}{v^2})^2$. From this, it follows easily that $tr\{\mathbf{I}_{12}^2(.)\}$ is the least for a nearly strongly balanced design. Further, for such a design, $\mathbf{C}(\tau|.)$ and $\mathbf{C}(\rho|.)$ are both completely symmetric. Hence the claim is established.

Theorem 6.4.2. For given (v,n,p), suppose there exists a nearly strongly balanced uniform **RMD** (v,n,p). Then it is universally optimal under both fixed- and mixed-effects models for the estimation of direct treatment effects among all competing designs which are uniform on units and on the last period.

Proof. We follow a simplified approach suggested by Magda (1980) and adopted by Kunert (1983), Mukhopahyay and Saha (1983) and Saharay (1986). Assume momentarily that the period effects are absent i.e., there are no differential period effects. Then, referring to (6.2.7)–(6.2.9) and (6.2.7)′–(6.2.9)′, it can be seen that for any competing design the expressions for $\mathbf{I}_{ij}(.)$'s are altered to

$$\left.\begin{aligned}
\mathbf{I}_{11}(F) &= \mathbf{r}^{\delta}-p^{-1}\mathbf{N}_u\mathbf{N}_u' = n\lambda_2(\mathbf{I}_v-\mathbf{J}_v/v),\\
\mathbf{I}_{22}(F) &= \bar{\mathbf{r}}^{\delta}-p^{-1}\mathbf{N}_u\mathbf{N}_u' = \{\lambda_1(p-1)-\frac{\lambda_1}{p}\}(\mathbf{I}_v-\frac{\mathbf{J}_v}{v}) + \frac{\lambda_1(p-1)}{pv}\mathbf{J}_v\\
\mathbf{I}_{12}(F) &= \mathbf{M}-p^{-1}\mathbf{N}_u\bar{\mathbf{N}}_u' = \mathbf{M}-n^{-1}\mathbf{r}\bar{\mathbf{r}}'\\
\mathbf{I}_{11}(M) &= (\mathbf{r}^{\delta}-n^{-1}p^{-1}\mathbf{r}\mathbf{r}') - \frac{w-\tilde{w}}{w}(p^{-1}\mathbf{N}_u\mathbf{N}_u'-n^{-1}p^{-1}\mathbf{r}\mathbf{r}')\\
&= n\lambda_2(\mathbf{I}_v-\mathbf{J}_v/v) = \mathbf{I}_{11}(F)\\
\mathbf{I}_{22}(M) &= (\bar{\mathbf{r}}^{\delta}-n^{-1}p^{-1}\mathbf{r}\mathbf{r}') - \frac{w-\tilde{w}}{w}(p^{-1}\bar{\mathbf{N}}_u\bar{\mathbf{N}}_u'-n^{-1}p^{-1}\bar{\mathbf{r}}\bar{\mathbf{r}}')\\
&= \{\lambda_1(p-1) - \frac{(w-\tilde{w})\lambda_1}{wp}\}\mathbf{I}_v-\{\frac{w-\tilde{w}}{w}(p-2)\lambda_1 + \frac{\tilde{w}}{w}\lambda_1(p-1)^2/p\}\mathbf{J}_v/v\\
\mathbf{I}_{12}(M) &= \mathbf{M}-n^{-1}p^{-1}\mathbf{r}\bar{\mathbf{r}}-\frac{w-\tilde{w}}{w}(p^{-1}\mathbf{N}_u\bar{\mathbf{N}}_u'-n^{-1}p^{-1}\mathbf{r}\bar{\mathbf{r}}')\\
&= \mathbf{I}_{12}(F) \;\; (\text{as } \mathbf{N}_u = \frac{p}{v}\mathbf{J} \text{ so that } \mathbf{N}_u\bar{\mathbf{N}}_u' = n^{-1}\mathbf{r}\bar{\mathbf{r}}')
\end{aligned}\right\} \qquad (6.4.3)$$

According to Ehrenfeld (1955), the information matrix (for any vector-parameter) under the *original* model (with period effects) is dominated by that under the *revised* model (without period effects) in the sense that the latter matrix minus the former matrix is nnd. This yields for any competing design which is uniform on units and on the last period,

$$\mathbf{C}(\tau|F) \le n\lambda_2(\mathbf{I}_v-\mathbf{J}_v/v)-\{\lambda_1(p-1) - \frac{\lambda_1}{p}\}^{-1}(\mathbf{M}-\frac{n(p-1)\mathbf{J}_v}{v^2})(\mathbf{M}-\frac{n(p-1)\mathbf{J}_v}{v^2})'$$

$$\mathbf{C}(\tau\,|M) \le n\lambda_2(\mathbf{I}_v - \mathbf{J}_v/v - (\mathbf{M} - \frac{n(p-1)\mathbf{J}_v}{v^2}[\{\lambda_1(p-1) - \frac{(w-\tilde{w})\lambda_1}{wp}\}\mathbf{I}_v - \{\frac{w-\tilde{w}}{w}(p-2)\lambda_1$$

$$+ \frac{\tilde{w}\lambda_1(p-1)^2}{wp}\}\mathbf{J}_v/v]^{-}(\mathbf{M} - \frac{n(p-1)\mathbf{J}_v}{v})'$$

with equality for the uniform design.

In either case, it is enough to show now that the nearly strongly balanced uniform **RMD** (v,n,p) minimizes both $tr[(\mathbf{M} - \frac{n(p-1)}{v^2}\mathbf{J}_v)(\mathbf{M} - \frac{n(p-1)}{v^2}\mathbf{J}_v)']$ and $tr[(\mathbf{M} - \frac{n(p-1)}{v^2}\mathbf{J}_v)[\{\lambda_1(p-1) - \frac{(w-\tilde{w})\lambda_1}{wp}\}\mathbf{I}_v - \{\frac{w-\tilde{w}}{w}(p-2)\lambda_1 + \frac{\tilde{w}\lambda_1(p-1)^2}{wp}\}\frac{\mathbf{J}_v}{v}]^{-}(\mathbf{M} - \frac{n(p-1)\mathbf{J}_v}{v^2})'\}$. The former has the expression $\sum_h\sum_{h'}\{m_{hh'} - \frac{n(p-1)}{v^2}\}^2 = \sum_h\sum_{h'}(m_{hh'} - npv^{-2} + A + Bv^{-1})^2$ for some A and B and the latter can be shown to be equal to $c\sum_h\sum_{h'}(m_{hh'} - \frac{n(p-1)}{v^2})^2 + d\sum_h\{s_h - \frac{n(p-1)}{v}\}^2$ for some $c>0$, $d>0$ where s_h is as defined in (6.2.6). The constraints on the $m_{hh'}$'s are $\sum_h(m_{hh'} - npv^{-2} + A + Bv^{-1}) = 0$ for every h'. (Vide (6.2.5) to see this). Moreover, $s_h = \sum_{h'=1}^{v} m_{hh'}$, $1 \le h \le v$. It is *not* difficult to see that the expression $\sum\sum(m_{hh'} - \frac{n(p-1)}{v^2})^2$ is made the least when (for every h') exactly B of the $m_{hh'}$'s are each equal to $npv^{-2} - A - 1$ and the rest $(v-B)$ of the $m_{hh'}$'s are each equal to $npv^{-2} - A$. This is precisely the case with the nearly strongly balanced uniform **RMD** (v,n,p) for which s_h's are also all equal to $\frac{n(p-1)}{v}$. Further, the relevant **C**-matrices are c.s.. Hence the result.

Theorem 6.4.3. For given (v,n,p), suppose there exists a nearly strongly balanced uniform **RMD** (v,n,p). Then it is universally optimal under both fixed- and mixed-effects models for the estimation of residual treatment effects among all competing designs which are uniform on units and on the first and last periods.

Proof. The basic approach is the same as above. In other words, we will work with a set-up where the period effects are absent so that (6.4.3) is relevant for the optimality study. However, Kunert (1983) has pointed out that whereas $\mathbf{I}_{11}(F)$ or $\mathbf{I}_{11}(M)$ in (6.4.3) has row- and column-sums zeroes, the matrices $\mathbf{I}_{22}(F)$ and $\mathbf{I}_{22}(M)$ in (6.4.3) do *not* enjoy this property. This implies that the relevant $\mathbf{C}(\rho\,|F)$ and $\mathbf{C}(\rho\,|M)$ matrices do *not* have zero row- and column-sums. Accordingly, Kiefer's Proposition 1 is *not* applicable as such. Kunert (1983) suggested a modification which implies that in such a situation, changing ρ to $\rho^* = (\mathbf{I}_v - \mathbf{J}_v/v)\rho$, one may work with revised information matrices for ρ^* which are of the form $(\mathbf{I}_v - \mathbf{J}_v/v)\mathbf{C}(\rho\,|.)(\mathbf{I}_v - \mathbf{J}_v/v)$. For any competing design, therefore, the revised information matrices are given by

$$\{\lambda_1(p-1) - \frac{\lambda_1}{p}\}(\mathbf{I}_v - \frac{\mathbf{J}_v}{v}) - (n\lambda_2)^{-1}\{\mathbf{M} - \frac{n(p-1)\mathbf{J}_v}{v^2}\}\{\mathbf{M} - \frac{n(p-1)\mathbf{J}_v}{v^2}\}'$$

and

$$\{\lambda_1(p-1) - \frac{(w-\tilde{w})\lambda_1}{wp}\}\{\mathbf{I}_v - \frac{\mathbf{J}_v}{v}\} - (n\lambda_2)^{-1}\{\mathbf{M} - \frac{n(p-1)\mathbf{J}_v}{v^2}\}\{\mathbf{M} - \frac{n(p-1)\mathbf{J}_v}{v^2}\}'.$$

In the above, we have used the fact that $(\mathbf{M} - \frac{n(p-1)\mathbf{J}_v}{v^2})\mathbf{J}_v = \mathbf{0}$ which is true this time as the competing designs are assumed to be *uniform* on the *first period* (which means that $s_h = \lambda_1(p-1),\ 1 \leq h \leq v$).

Once more, it is now straightforward to verify that a nearly strongly balanced uniform **RMD** maximizes the trace of each of the above matrices.

Remark 6.4.1. Kunert (1983) tried to generalize these results in the sense of extending the class of competing designs to a wider class. For the direct effects, he succeeded in showing that a nearly strongly balanced uniform **RMD** (v,n,p) is universally optimal over the set of all designs under a fixed-effects non-circular model whenever $Av \geq B(v-B-1)$ and $p/v \geq \max(2, \frac{B(v-B)}{4} + \frac{2}{v})$ where $n = Av^2+Bv$.

5. **Universal Optimality of Balanced Uniform RMDs**

In this section we present some results on universal optimality of balanced uniform **RMDs** over the subclass of designs where no treatment is allowed to precede itself over the periods on any unit whatsoever. Recall the definition of a balanced design as given in Definition 4. In terms of the **M**-matrix, for every design in this subclass under consideration, $m_{hh} = 0$ for all $1 \leq h \leq v$ while for the balanced design, we have additionally $m_{hh'} = \lambda$ (say), for all $1 \leq h \neq h' \leq v$.

For a balanced uniform **RMD** $(v, n = \lambda_1 v, p = \lambda_2 v)$, *under a non-circular model*

$$\left.\begin{aligned} &\mathbf{I}_{11}(F) = \mathbf{I}_{11}(M) = \lambda_1 p(\mathbf{I}_v - \mathbf{J}_v/v) \\ &\mathbf{I}_{12}(F) = \mathbf{I}_{12}(M) = -\lambda(\mathbf{I}_v - \mathbf{J}_v/v),\ \lambda = \lambda_1(p-1)/(v-1) \\ &\mathbf{I}_{22}(F) = \lambda_1(p-1-\frac{1}{p})(\mathbf{I}_v - \mathbf{J}_v/v) \\ &\mathbf{I}_{22}(M) = \lambda_1(p-1-\frac{w-\tilde{w}}{wp})(\mathbf{I}_v - \mathbf{J}_v/v) \end{aligned}\right\} \tag{6.5.1}$$

while *under a circular model*

$$\left.\begin{aligned} &\mathbf{I}_{11}(F) = \mathbf{I}_{22}(F) = \mathbf{I}_{11}(M) = \mathbf{I}_{22}(M) = \lambda_1 p(\mathbf{I}_v - \mathbf{J}_v/v) \\ &\text{and } \mathbf{I}_{12}(M) = \mathbf{I}_{12}(F) = (-\lambda_1 p/(v-1))(\mathbf{I}_v - \mathbf{J}_v/v) \end{aligned}\right\} \tag{6.5.2}$$

We enunciate the following results. The fixed-effects part is due to Cheng and Wu (1980) and Magda (1980), and the mixed-effects part is due to Mukhopadhyay and Saha (1983) and Saharay (1986).

Theorem 6.5.1. For given $v,\ n = \lambda_1 v,\ p = \lambda_2 v$, a balanced uniform design is universally optimal for the estimation of direct effects in a non-circular fixed- or mixed-effects model among all **RMDs** for which $m_{hh} = 0$ for all $1 \leq h \leq v$ and which are uniform on each unit and the last period. In a circular fixed- or mixed-effects model, the competing **RMDs** have to satisfy only the condition $m_{hh} = 0$ for all $1 \leq h \leq v$.

Theorem 6.5.2. For given v, $n = \lambda_1 v$ and $p = \lambda_2 v$, $(\lambda_2 \geq 2)$, a balanced uniform design is universally optimal for the estimation of residual effects in a non-circular fixed- or mixed-effects model among all **RMDs** for which $m_{hh} = 0$ for all $1 \leq h \leq v$ and which are equireplicate in the first $(p-1)$ periods (so that $\tilde{r}_h = \lambda_1(p-1)$ for all $1 \leq h \leq v$). In a circular fixed- or mixed-effects model, the competing **RMDs** have to satisfy only the condition $m_{hh} = 0$ for all $1 \leq h \leq v$.

Theorem 6.5.3. For given v, $n = \lambda_1 v$ and $p = v$, a balanced uniform design is universally optimal for the estimation of residual effects in a circular or non-circular fixed- or mixed-effects model among all **RMDs** for which $m_{hh} = 0$ for all $1 \leq h \leq v$.

As regards the non-circular model, referring to (6.5.1), it follows that for a balanced uniform design, $\mathbf{C}(\tau|.)$ and $\mathbf{C}(\rho|.)$ are completely symmetric. Thus trace maximization is to be verified next. In the case of residual effects, for any competing design, the relevant **C**-matrix $\mathbf{I}_{22} - \mathbf{I}_{21}\mathbf{I}_{11}^{-}\mathbf{I}_{12}$ is dominated by $\mathbf{I}_{22} - \mathbf{I}_{21}\mathbf{I}_{12}(\lambda_1 p)^{-1}$ as $\mathbf{I}_{11}^{-} \geq (\lambda_1 p)^{-1}\mathbf{I}_v$ by a result of Wu (1980). Further, in the above, equality holds for the balanced uniform **RMD**.

On the other hand, $\mathbf{I}_{22} \geq (\lambda_1(p-1))^{-1}\mathbf{I}_v$ but equality does *not* occur even for the balanced uniform **RMD**. Thus, for the case of direct effects, the computation of $tr(\mathbf{I}_{11} - \mathbf{I}_{12}\mathbf{I}_{22}^{-}\mathbf{I}_{21})$ gets complicated. The conditions of uniformity on each unit and on the last period are brought in to facilitate the exercise on trace maximization.

We refer to Cheng and Wu (1980, 1983) for the details in both the cases. For the circular model, we refer to Magda (1980).

The following results are to be found in Mukhopadhyaya and Saha (1983) and Saharay (1986) regarding optimality of balanced uniform **RMDs** with $p = \lambda_2 v + 1$ in a non-circular fixed- or mixed-effects model.

Theorem 6.5.4. For given v, $n = \lambda_1 v$, $p = \lambda_2 v + 1$, $v \geq 3$, a design which is balanced, uniform on each period, and uniform on each unit in the first $(p-1)$ periods is universally optimal for the estimation of direct effects in a non-circular fixed- or mixed-effects model over the class of designs for which $m_{hh} = 0$ for all $1 \leq h \leq v$ and which are equireplicate and uniform in the last period.

Theorem 6.5.5. For given v, $n = \lambda_1 v$, $p = \lambda_2 v + 1$, $v \geq 3$, a design which is balanced, uniform on each unit in the first $(p-1)$ periods and uniform on each period is universally optimal for the estimation of residual effects in a non-circular fixed- or mixed-effects model over the class of designs for which $m_{hh} = 0$ for all $1 \leq h \leq v$ and which are uniform on each unit in the first $(p-1)$ periods and also uniform in the last period.

6. Concluding Remarks

It is surprising that almost the entire available literature on optimal **RMDs** deal with a set-up where $n = \lambda_1 v$ and $p = \lambda_2 v$ for some λ_1, $\lambda_2 \geq 1$. The only exception is Dey *et al* (1983) where it is assumed that $p < v$. Even for moderate values of v, the assumptions that both n and p are integral multiples of v may *not* be tenable. Certainly, more work need be done specifically in cases of $p \ll v$ with the units chosen at random. Of course, it may *not* be possible to obtain universally optimal designs in many cases. However, designs optimal w.r.t. some specific criteria might as well be preferred from practical considerations.

For ready reference, we now display most of the optimality results in a tabular form. Before that, we give some examples of optimal **RMDs**.

Examples of universally optimal designs

1. Strongly Balanced **URMD** $(v = 3, n = 9, p = 6)$

		units								
		1	2	3	4	5	6	7	8	9
	1	1	1	1	2	2	2	3	3	3
	2	1	2	3	1	2	3	1	2	3
periods	3	2	2	2	3	3	3	1	1	1
	4	2	3	1	2	3	1	2	3	1
	5	3	3	3	1	1	1	2	2	2
	6	3	1	2	3	1	2	3	1	2

2. Strongly Balanced **RMD** $(v = 3, n = 9, p = 7)$ uniform on periods and uniform on units in the first $(p-1)$ periods.

		units								
		1	2	3	4	5	6	7	8	9
	1	1	1	1	2	2	2	3	3	3
	2	1	2	3	1	2	3	1	2	3
periods	3	2	2	2	3	3	3	1	1	1
	4	2	3	1	2	3	1	2	3	1
	5	3	3	3	1	1	1	2	2	2
	6	3	1	2	3	1	2	3	1	2
	7	1	1	1	2	2	2	3	3	3

3. **URMD** $(v = 3, n = 9, p = 6)$ with $\mathbf{M} = \mathbf{1}\tilde{\mathbf{r}}'/v$ and having preperiod treatments with a non-circular arrangement.

		units								
		1	2	3	4	5	6	7	8	9
pre-period	0	1	1	1	1	1	1	1	1	1
	1	1	1	1	2	2	2	3	3	3
	2	1	2	3	1	2	3	1	2	3
periods	3	2	2	2	3	3	3	1	1	1
	4	2	3	1	2	3	1	2	3	1
	5	3	3	3	1	1	1	2	2	2
	6	3	1	2	3	1	2	3	1	2

4. Nearly Strongly Balanced **URMD** ($v = 3, n = p = 6$)

		Units					
		1	2	3	4	5	6
	1	1	2	3	1	2	3
	2	2	3	1	1	2	3
periods	3	3	1	2	2	3	1
	4	3	1	2	3	1	2
	5	2	3	1	3	1	2
	6	1	2	3	2	3	1

5. Balanced **URMD** ($v = 3, n = 12, p = 3$)

		Units											
		1	2	3	4	5	6	7	8	9	10	11	12
	1	1	2	3	2	3	1	1	2	3	2	3	1
periods	2	3	1	2	3	1	2	3	1	2	3	1	2
	3	2	3	1	1	2	3	2	3	1	1	2	3

6. Balanced **RMD** ($v = 3, n = 12, p = 4$) uniform on periods and uniform on each unit in the first $(p-1)$ periods.

		Units											
		1	2	3	4	5	6	7	8	9	10	11	12
	1	1	2	3	2	3	1	1	2	3	2	3	1
periods	2	3	1	2	3	1	2	3	1	2	3	1	2
	3	2	3	1	1	2	3	2	3	1	1	2	3
	4	1	1	2	3	1	1	3	2	2	3	3	2

Sl. No.	Design Parameters	Restrictions on the class of competing designs	Model(s) specification	Nature of universally optimal designs	Reference
1.	$v, n = \lambda_1 v, p = \lambda_2 v$	NIL	Circular/Non-circular Fixed-/Mixed-effects	Strongly Balanced **URMD** (Direct and Residual Effects)	Theorem 6.3.1
2.	$v, n = \lambda_1 v, p = \lambda_2 v + 1$	NIL	Non-circular Fixed-/Mixed-effects	Strongly Balanced **RMD** uniform on periods uniform on units in the first $(p-1)$ periods (Direct and Residual Effects)	Theorem 6.3.2
3.	$v, n = \lambda_1 v, p = \lambda_2 v$	NIL	Pre-period with arbitrary arrangement of treatments Fixed-/Mixed-effects	**URMD** with $\mathbf{M} = \mathbf{1}\bar{\mathbf{r}}'/v$ (Direct effects)	Theorem 6.3.5
4.	$v, n = \lambda_1 v, p = \lambda_2 v$	Uniform on periods and units	Non-circular Fixed-/Mixed-effects	Nearly strongly balanced **URMD** (Direct and Residual effects)	Theorem 6.4.1
5.	$v, n = \lambda_1 v, p = \lambda_2 v$	Uniform on units and the last period	Non-circular Fixed-/Mixed-effects	Nearly strongly balanced **URMD** (Direct effects)	Theorem 6.4.2
6.	$v, n = \lambda_1 v, p = \lambda_2 v$	Uniform on units and on the first and last periods	Non-circular Fixed-/Mixed-effects	Nearly strongly balanced **URMD** (Residual effects)	Theorem 6.4.3
7.	$v, n = \lambda_1 v,$ $p = \lambda_2$	No treatment precedes itself; uniform on each unit and the last period	Non-circular Fixed-/Mixed-effects	Balanced **URMD** (Direct effects)	Theorem 6.5.1
8.	$v, n = \lambda_1 v,$ $p = \lambda_2 v$	No treatment precedes itself	Circular Fixed-/Mixed-effects	Balanced **URMD** (Direct effects)	Theorem 6.5.1

9.	$v, n = \lambda_1 v$, $p = \lambda_2 v, \lambda_2 \geq 2$	No treatment precedes itself; equireplicate in the first $(p-1)$ periods	Non-circular Fixed-/Mixed-effects	Balanced **URMD** (Residual effects)	Theorem 6.5.2
10.	$v, n = \lambda_1 v$, $p = \lambda_2 v, \lambda_2 \geq 2$	No treatment precedes itself	Circular Fixed-/Mixed-effects	Balanced **URMD** (Residual effects)	Theorem 6.5.2
11.	$v, n = \lambda_1 v$, $p = v$	No treatment precedes itself	Circular/Non-circular Fixed-/Mixed-effects	Balanced **URMD** (Residual effects)	Theorem 6.5.3
12.	$v, n = \lambda_1 v$, $p = \lambda_2 v + 1$, $v \geq 3$	No treatment precedes itself; equireplicate and uniform in the last period	Non-circular Fixed-/Mixed-effects	Balanced **RMD** uniform on periods and uniform on each unit in the first $(p-1)$ periods (Direct effects)	Theorem 6.5.4
13.	$v, n = \lambda_1 v$, $p = \lambda_2 v + 1, v \geq 3$	No treatment precies itself; uniform on each unit in the first $(p-1)$ periods and uniform in the last period	Non-circular Fixed-/Mixed-effects	Balanced **RMD** uniform on periods and uniform on each unit in the first $(p-1)$ periods (Residual effects)	Theorem 6.5.5

REFERENCES

Afsarinejad, K. (1983). Balanced repeated measurements designs. *Biometrika, 70*, 199-204.

Azzalini, A. and Giovagnoli, A. (1987). Some optimal designs for repeated measurements with autoregressive errors. *Biometrika, 74*, 725-734.

Cheng, C.S. and Wu, C.F. (1980). Balanced repeated measurements designs. *Ann. Statist., 8*, 1272-1283.

Cheng, C.S. and Wu, C.F. (1983). Corrections to balanced repeated measurements designs. *Ann. Statist., 11*, 349.

Constantine, G. and Hedayat, A. (1979). Repeated measurements designs III. Tech. Report, Univ. Illinois at Chicago.

Constantine, G. and Hedayat, A. (1982). A construction of repeated measurements designs with balance for residual effects. *J. Statist. Planning and Inference, 6*, 153-164.

Dey, A., Gupta, V.K. and Singh, M. (1983). Optimal change-over designs. *Sankhyā (B), 45*, 233-239.

Ehrenfeld, S. (1955). Complete class theorems in experimental designs. *Proceedings of Third Berkeley Symposium*, Univ. California Press, **1**, 57-67..

Finney, D.J. and Outhwaite, A.D. (1955). Serially balanced sequences. *Nature, 176*, 748.

Finney, D.J. and Outhwaite, A.D. (1956). Serially balanced sequences in bioassay. *Proc. Roy. Soc., B, 145*, 493-507.

Gill, P.S. and Shukla, G.K. (1987). Optimal change-over designs for correlated observations. *Comm. Statist. (A) Theo. Meth., 16*, 2243-2262.

Hedayat, A. and Afsarinejad, K. (1975). Repeated measurements designs I. In *A Survey of Statistical Designs and Linear Models*, Edtd. Srivastava, 229-242. North-Holland.

Hedayat, A. and Afsarinejad, K. (1978). Repeated measurements designs II. *Ann. Statist., 6*, 619-628.

Hedayat, A. (1981). Repeated measurements designs IV. Recent advances, *Bull. Inter. Statist.* Proceedings of the 43rd Session, XLIX, Book 1, 591-605.

Kiefer, J. (1975). Construction and optimality of generalized Youden designs. In a *Survey of Statistical Design and Linear Models*. (J.N. Srivastava, ed.). Amsterdam: North-Holland Publishing Co., (1975), 333-353.

Kunert, J. (1983). Optimal design and refinement of the linear model with application to repeated measurements designs. *Ann. Statist.*, *11*, 247-257.

Kunert, J. (1984a). Optimality of balanced uniform repeated measurements designs. *Ann. Statist.*, *12*, 1006-1017.

Kunert, J. (1984b). Designs balanced for circular residual effects. *Comm. Statist. (A) Theo. Meth.* *21*, 2665-2671.

Kunert, J. (1985). Optimal repeated measurements designs for correlated observations and analysis by weighted least squares. *Biometrika*, *72*, 375-389.

Magda, G.C. (1980). Circular balanced repeated measurements designs. *Comm. Statist. (A) Theo. Meth.* *9*, 1901-1918.

Mathews, J.N.S. (1987). Optimal cross-over designs for the comparison of two treatments in the presence of carryover effects and autocorrelated errors. *Biometrika*, *74*, 311-320.

Mukerjee, R. and Sen, M. (1985). Universal optimality of a class of type 2 and allied sequences. *Sankhyā (B)*, *47*, 216-223.

Mukhopadhyay, A.C. and Saha, R. (1983). Repeated measurements designs. *Cal. Statist. Assoc. Bull.*, *32*, 153-168.

Pigeon, J.G. and Raghavarao, D. (1987). Cross-over designs for comparing treatments with a control. *Biometrika*, *74*, 321-328.

Roy, B.K. (1988). Construction of strongly balanced uniform repeated measurements designs. *J. Statist. Planning and Inference*, *19*, 341-348.

Saharay, R. (1986). *Studies on Optimality of Some Classes of Designs*. Unpublished Ph.D. Thesis. Indian Statistical Institute, Calcutta, India.

Sampford, M.R. (1957). Methods of construction and analysis of serially balanced sequences. *J. Roy. Statist.* (B), *19*, 286-304.

Sen, M. and Mukerjee, R. (1987). Optimal repeated measurements designs with interaction. *J. Statist. Planning and Inference*, *17*, 81-92.

Sen, M. and Sinha, B.K. (1986). A statistical analysis of serially balanced sequences: First order residuals proportional to direct effects. *Cal. Statist. Assoc. Bull*, *35*, 31-36.

Street, D.J. (1988). Some repeated measurements designs. *Comm. Statist. (A) Theo. Meth.*, *17*, 87-104.

Sinha, B.K. (1975). On some optimum properties of serially balanced sequences. *Sankhyā (B)*, *37*, 173-192.

Williams, E.R. (1987). A note on change-over designs. *Austr. J. Statist., 29*, 309-316.

Wu, C.F. (1980). On some ordering properties of the generalized inverses of nonnegative definite matrices. *Linear Algebra and Its Applications, 32*, 49-60.

CHAPTER SEVEN

OPTIMAL DESIGNS FOR SOME SPECIAL CASES

1. Introduction

In the previous Chapters we discussed optimal designs under standard models and with the optimality functionals symmetric in all the treatments. In this Chapter we discuss some situations where some of these considerations do not apply. In Section 7.2 we deal with optimality aspects of designs when the observations have a covariance pattern. Designs where we have one or more covariates are discussed in Section 7.3. Finally in Section 7.4 we present some results on optimal designs for comparing a set of treatments against a control treatment.

2. Models with Correlated Observations

When the observations are correlated, the covariance structure plays an important role both in the analysis of a given design as well as in the choice of an efficient design. If the covariance structure is *precisely* known, one can look for designs which are optimal when the weighted least squares (**WLS**) estimates based on this error structure are used. If, on the other hand, the covariance structure is not precisely known one cannot construct **WLS** estimates. The following two stage procedure due to Kiefer and Wynn (1981) appears to be very appealing in such situations especially if the departure from the standard covariance structure is small. At stage one, we look for a class of designs which are optimal w.r.t. the standard covariance structure. At the second stage we consider the ordinary least squares (**OLS**) estimates and look for designs within the class obtained at stage one which maximize the precision of these **OLS** estimates w.r.t. the true covariance structure. This approach often yields designs which are optimal w.r.t. a wide class of covariance structures and leads to some interesting designs which are also useful in some other contexts. In this section we shall develop results along these lines for block designs and for row-column designs.

Following Kiefer and Wynn (1981), we write

$$E(\mathbf{Y}) = \mathbf{X}\boldsymbol{\theta}, \qquad Cov(\mathbf{Y}) = \mathbf{V} \tag{7.2.1}$$

where $\mathbf{Y}$ is the $n\times 1$ vector of observations, $\mathbf{X}$ is the $n\times p$ design matrix, $\boldsymbol{\theta}$ is the $p\times 1$ vector of parameters and $\mathbf{V}$ is the $n\times n$ matrix which specifies the true covariance structure. Standard covariance structure is given by $\mathbf{V}_0 = \mathbf{I}_n\sigma^2$. Estimable parametric functions may be written as $\boldsymbol{\gamma} = \mathbf{B}\boldsymbol{\theta}$. The **WLS** estimator of $\boldsymbol{\gamma}$ under (7.2.1) is given by

$$\hat{\boldsymbol{\gamma}}_{\mathbf{V}} = \mathbf{B}(\mathbf{X}'\mathbf{V}^{-1}\mathbf{X})^{+}\mathbf{X}'\mathbf{V}^{-1}\mathbf{Y} \tag{7.2.2}$$

where, as before, $\mathbf{A}^{+}$ denotes the Moore-Penrose inverse of a matrix $\mathbf{A}$. When $\mathbf{V} = \mathbf{V}_0$, the corresponding estimate is given by

$$\hat{\gamma}_0 = \hat{\gamma}_{V_0} = \mathbf{B}(\mathbf{X}'\mathbf{X})^{+}\mathbf{X}'\mathbf{Y}.$$

We shall first deal with the case of block designs. For a block design, $\mathbf{X}$ is of order $bk \times (b+v+1)$ and the elements of γ are the standard treatment contrasts $\tau_i - (\sum \tau_i / v)$. Results of Chapter Two show that a **BIBD** is universally optimal in the extended sense when $\mathbf{V} = \mathbf{V}_0$.

For the second stage described above, we consider the set of $\mathbf{D}$–matrices

$$\mathbf{D}(\mathbf{X},\mathbf{V}) = Cov(\hat{\gamma}_0 | \mathbf{V}) \tag{7.2.3}$$

where $\mathbf{X}$ is a design matrix for a **BIBD** and $Cov(\cdot|\mathbf{V})$ indicates the dispersion matrix when $\mathbf{V}$ is the true covariance structure. Since we are working with dispersion matrices, it seems natural to minimize an appropriate functional of $\mathbf{D}$. (Recall that A–, D– and E–optimality criteria can be expressed directly as functionals of $\mathbf{D}$). Again, as γ refers to the vector of standard treatment contrasts, we consider the class of permutation invariant convex functionals ψ which are non-decreasing i.e. which satisfy $\psi(b\mathbf{D}) \geq \psi(\mathbf{D})$ $\forall$ $b>1$. A design minimizing $\psi(\mathbf{D})$ for *all* such ψ has been called *weakly universally optimal* (**WUO**) by Kiefer and Wynn. They have pointed out that the class of such criteria includes A–, E– and ϕ_p optimality but does *not* include D–optimality.

Let $\mathbf{Q}$ denote the vector of adjusted treatment totals defined by

$$kQ_i = kT_i - \sum_j n_{ij} B_j$$

where T_i denotes the total yield of the i-th treatment and B_j denotes the total yield for the j-th block and n_{ij} is the (i,j)th element of the incidence matrix $\mathbf{N}$. We also define $k \times v$ matrices $\mathbf{P}_j$ where

$$\{\mathbf{P}_j\}_{li} = \begin{cases} 1 & \text{if the } l\text{-th plot in the } j\text{-th} \\ & \text{block receives the } i\text{-th treatment} \\ 0 & \text{otherwise.} \end{cases}$$

Clearly, the matrices $\mathbf{P}_j$ are determined by the design matrix $\mathbf{X}$. It is easy to verify that

$$\mathbf{Q} = \sum_j \mathbf{P}_j'(\mathbf{I}_k - \mathbf{J}_{kk}/k)\mathbf{Y}_j$$

where $\mathbf{Y}_j$ is the $k \times 1$ vector of observations in the j-th block. Since for a **BIBD** $\hat{\gamma}_0 = k\mathbf{Q}/\lambda v$, we get, for an alternative variance structure $\mathbf{V}^* = \Delta \otimes \mathbf{I}_b$,

$$\mathbf{D}(\mathbf{X},\mathbf{V}^*) = (k^2/\lambda^2 v^2)Cov(\mathbf{Q}) = (k^2/\lambda^2 v^2)\sum_j \mathbf{P}_j'\mathbf{W}\mathbf{P}_j$$

where

$$\mathbf{W} = (\mathbf{I}_k - \mathbf{J}_{kk}/k)\Delta(\mathbf{I}_k - \mathbf{J}_{kk}/k).$$

Before we attempt to find weakly universally optimal designs, we note that since $\mathbf{P}_j\mathbf{P}_j' = \mathbf{I}_k$,

$$tr(\mathbf{D}(\mathbf{X},\mathbf{V}^*)) = (bk^2/\lambda^2 v^2)tr(\mathbf{W}).$$

Thus, $tr(\mathbf{D}(\mathbf{X},\mathbf{V}^*))$ is the same for all **BIBDs** and hence every **BIBD** is equally good in terms of A–optimality for *every* $\mathbf{V}^*$ if one uses $\hat{\gamma}_0$ to estimate γ.

Regarding other functionals it is easy to see that a **BIBD** with $\mathbf{X} = \mathbf{X}^*$ is **WUO** w.r.t. the structure $\mathbf{V}^*$ if
(i) $\mathbf{D}(\mathbf{X}^*,\mathbf{V}^*)$ is completely symmetric
and (ii) $tr(\mathbf{D}(\mathbf{X}^*,\mathbf{V}^*)) = \min tr(\mathbf{D}(\mathbf{X},\mathbf{V}^*))$
where the minimum is taken over $\mathbf{X}$ for all **BIBDs**.

We shall now consider specific structures for $\mathbf{V}^*$. Kiefer and Wynn (1981) studied the nearest neighbor (**NN**) covariance structure defined by $\Delta = ((\rho_{rs}))\sigma^2$ where

$$\rho_{rs} = \begin{cases} 1 & \text{if } r = s \\ \rho & \text{if } |r-s| = 1 \\ 0 & \text{otherwise.} \end{cases}$$

From the covariance matrix of $\mathbf{Q}$ described above, it can be deduced that for the **NN** structure

$$k^2\sigma^{-2}Var(Q_i) = r[k(k-1)+2\rho(k+1)] + 2\rho ke_i$$

and

$$k^2\sigma^{-2}Cov(Q_i,Q_{i'}) = -\lambda[k+2\rho(k+1)] + k\rho[2f_{ii'}+e_{ii'}+kN_{ii'}], \quad i \neq i'$$

where,

e_i = # of blocks with treatment i at an end

$N_{ii'}$ = # of blocks with treatments i and i' in adjacent positions

$e_{ii'}$ = # of blocks with treatments i and i' but only *one* occurring at an end and $f_{ii'}$ = # of blocks with both treatments i and i' at the end positions.

Since $\sum_{i'(\neq i)} (2f_{ii'}+e_{ii'}+kN_{ii'}) = 2r(k+1)-2e_i$, a sufficient condition for **WUO** is that

$$2f_{ii'}+e_{ii'}+kN_{ii'} \text{ is the same for all pairs } (i,i'). \tag{7.2.4}$$

When this condition holds, dividing $\sum_{i\neq i'}(2f_{ii'}+e_{ii'}+kN_{ii'})$ by $v(v-1)/2$ gives the condition that 4λ is divisible by k. In particular, if $k = 3$, λ must be a multiple of 3. Now, for a **BIBD** with $k = 3$ and $\lambda = 3m$, the parameters must be of the form

$$k = 3, \lambda = 3m, b = mv(v-1)/2, r = 3m(v-1)/2.$$

Kiefer and Wynn (1981) have proved the existence of **BIBDs** with these parameters for which condition (7.2.4) holds. Cheng (1983) gave methods for construction of such designs. An example of such a design with $b = 21$, $v = 7$, $r = 9$, $k = 3$, $\lambda = 3$ is given by

1 2 4	1 2 6	1 4 2
2 3 5	2 7 3	4 3 5
3 4 6	3 6 4	5 2 6
4 5 7	6 5 7	2 3 7
5 6 1	7 4 1	3 6 1
6 7 2	4 5 2	6 7 4
7 1 3	5 1 3	7 1 5

For this design $f_{ii'} = 1$ and $N_{ii'} = 2$ for all $i \neq i'$. It should be noted that in the above example, every treatment occurs precisely 3 times in each position. Even though this feature is not required by the model, it would be advisable to retain it.

When $k = 4$, Hanani (1961) has shown that the usual necessary conditions for the existence of a **BIBD** are also sufficient. However, Kiefer and Wynn (1981) have shown that designs with condition (7.2.4) do not always exist.

Russell and Eccleston (1987) have given an algorithm which for every **BIBD** gives a re-arrangement of treatments among the plots within each block in such a way that $(2f_{ii'}+e_{ii'}+N_{ii'})$'s are as nearly equal as possible.

For large k the contribution of $N_{ii'}$ to $2f_{ii'}+e_{ii'}+kN_{ii'}$ will be dominant and hence it would be useful to consider **BIBDs** for which $N_{ii'}$'s are all equal. Such **BIBDs** are termed equineighbored **BIBDs** (**EBIBDs**) by Kiefer and Wynn. For $k = 3$ it is easy to see that the equality of the $N_{ii'}$'s is enough for condition (7.2.4) to hold. Cheng (1983) has given methods of construction of **EBIBDs** for $k = v-2$ and $k = v-1$.

Another structure of interest known as autoregressive structure is one for which

$$\rho_{rs} = \rho^{|r-s|}.$$

For $k = 3$ analysis analogous to that for the **NN** structure shows that a **BIBD** is **WUO** w.r.t. the autoregressive structure if $f_{ii'}$ is the same for all $i \neq i'$. Since $f_{ii'} + N_{ii'} = \lambda$ for $k = 3$, it turns out that an **EBIBD** is **WUO** for both the structures. For higher values of k, the construction of **WUO** designs w.r.t. the autoregressive structure appears to be very cumbersome.

Recently Gill and Shukla (1985) have considered analysis of designs based on the use of $\hat{\gamma}_{\mathbf{V}}$ as opposed to $\hat{\gamma}_{\mathbf{V}_0}$. They have looked at some **BIBDs** with $k = v$ for which $N_{ii'}$'s are all equal and $f_{ii'}$'s take only the values 0 and 1. They computed the **A**– and **D**–efficiencies of such designs with $v = 5$ and $v = 10$ for each of the two structures described above. Their **A**– and **D**–efficiencies are measures of departure from complete symmetry for the **C**–matrix. For the autoregressive structure they find that the efficiencies are nearly one i.e. the **C**–matrices are nearly completely symmetric.

We note that the use of $\hat{\gamma}_{\mathbf{V}}$ is feasible only when $\mathbf{V}$ is known or when a good estimate of $\mathbf{V}$ is available. Analysis of a design based on estimated $\mathbf{V}$ tends to overestimate the actual precision and hence comparisons with **OLS** estimates can be somewhat misleading. On the other hand Kiefer and Wynn approach is a good one only when the departures from the standard covariance structure are relatively small w.r.t the usual Euclidean norm. Thus, the two approaches appear to complement each other.

Kiefer and Wynn (1981) have also examined the nearest neighbor structure for row-column designs. They have considered designs in v rows and v columns for comparing v treatments where the covariance between two observations is $\rho\sigma^2$ if they are adjacent in a row or in a column and is zero otherwise. Again at the first stage the Latin Square design is extended universally optimal. Also, all Latin Squares are equally good w.r.t. the **A**–optimality criterion for the **NN** structure. Condition analogous to (7.2.4) for a Latin Square design to be **WUO** for all values of ρ is that every pair of treatments occurs in adjacent position precisely four times. Kiefer and Wynn have also stated that a Latin Square in which the (j,k)-th cell contains treatment number

$$\{\sum_{r=1}^{j}(-1)^r(r-1) + \sum_{r=1}^{k}(-1)^r(r-1)\} \mod v$$

has this property. For $v = 5$ we get the following square

5	1	4	2	3
1	2	5	3	4
4	5	3	1	2
2	3	1	4	5
3	4	2	5	1

We would like to remark that there are many covariance structures of interest both in block designs and in row-column designs. As we have seen in the above special cases, search for optimal designs in such structures leads to challenging combinatorial problems. It may also be remarked that the above analysis precludes the use of any randomization other than re-labelling of treatments.

3. Models with Covariates

We shall consider designs without blocking for comparing v treatments in the presence of k covariates. As usual, our primary interest is in comparing the treatments. However, we shall also consider the optimality aspects of estimation of regression coefficients for the covariates.

We consider the case where we have covariate values z_{im} for the i-th experimental unit (e.u.) for $m = 1,2, \ldots, k$; $i = 1,2, \ldots, N$. We first consider the allocation of the experimental units to the v treatments where the j-th treatment is applied on n_j units, the n_j's being fixed. Thus, the model is

$$y_i = \mu + \sum_{j=1}^{v} \tau_j x_{ij} + \sum_{m=1}^{k} \beta_m z_{im} + \epsilon_i \tag{7.3.1}$$

where μ denotes the general mean, τ_j denotes the effect of the j-th treatment, β_m denotes the effect of the m-th covariate, $x_{ij} = 1(0)$ if the i-th e.u. receives (does not receive) the j-th treatment and ϵ_i's are the errors assumed to be uncorrelated with zero expectations and to have a common variance.

A more common and meaningful situation is one in which the z_{im}'s are fixed i.e. are not under experimenter's control. Designs for this situation are considered by Haggstrom (1975), Harville (1975) and Wu (1981). However, there may be situations where the z_{im}'s are under the experimenter's control. Optimal designs for this later situation are considered by Lopes Troya (1982a, 1982b).

Following Haggstrom (1975) we introduce the following definition of balance for a design with covariates.

Definition 7.3.1: A design is said to be balanced if $\bar{z}_{jm} = \sum_{i=1}^{N} z_{im} x_{ij} / n_j$ is independent of j for $m = 1,2, \ldots, k$.

Next we work out the expression for the joint information matrix of (τ, β) in the usual way. It can be easily verified that this assumes the form

$$I\binom{\tau}{\beta} = \begin{bmatrix} \mathbf{n}^{\delta} - \dfrac{\mathbf{n}\mathbf{n}'}{N} & ((n_j(\bar{z}_{jm} - \bar{z}_{.m}))) \\ & \mathbf{S}_T \end{bmatrix} = \begin{bmatrix} \mathbf{n}^{\delta} - \dfrac{\mathbf{n}\mathbf{n}'}{N} & \Delta \\ & \mathbf{S}_T \end{bmatrix} \tag{7.3.2}$$

where $\mathbf{n}^{\delta} = Diag(n_1, \ldots, n_v)$, $\mathbf{n} = (n_1, \ldots, n_v)'$, $\bar{z}_{.m} = \sum_{i=1}^{N} z_{im}/N$ and $\mathbf{S}_T = ((S_T(r,s))) =$ matrix of total S.S. and S.P. with

$$S_T(r,s) = \sum_{i=1}^{N} z_{ir} z_{is} - N\bar{z}_{.r}\bar{z}_{.s}, \quad 1 \le r,\ s \le k.$$

Next note that $\mathbf{S}_T = \mathbf{S}_W + \mathbf{S}_B$ where $\mathbf{S}_W = ((S_W(r,s)))$ with $S_W(r,s) = \sum_{i=1}^{N} z_{ir} z_{is} - \sum_{j=1}^{v} n_j \bar{z}_{jr} \bar{z}_{js}$,

$1 \le r,s \le k$. and $\mathbf{S}_B = ((S_B(r,s)))$ with $\mathbf{S}_B(r,s) = \sum_{j=1}^{v} n_j(\bar{z}_{jr}\bar{z}_{js} - \bar{z}_{.r}\bar{z}_{.s})$, $1 < r,s \le k$. From (7.3.2) it follows that

$$\mathbf{I}(\boldsymbol{\beta}) = \mathbf{S}_W. \tag{7.3.3}$$

For a balanced design, $\bar{z}_{jm} = \bar{z}_{.m}$ for all $j = 1,2, \ldots, v$ and this is true for every m, $1 \le m \le k$. This leads to Δ and $\mathbf{S_B}$ being null matrices of orders $v \times k$ and $v \times v$ respectively. Therefore, the above form of the information matrix reduces to

$$\mathbf{I}_0\binom{\boldsymbol{\tau}}{\boldsymbol{\beta}} = \begin{bmatrix} \mathbf{n}^\delta - \dfrac{\mathbf{nn}'}{N} & 0 \\ 0 & \mathbf{S}_T \end{bmatrix} \tag{7.3.4}$$

whence

$$\mathbf{I}_0(\boldsymbol{\tau}) = \mathbf{n}^\delta - \frac{\mathbf{nn}'}{N} \text{ and } \mathbf{I}_0(\boldsymbol{\beta}) = \mathbf{S}_T. \tag{7.3.5}$$

In view of the above, we now have the following strong optimality results w.r.t. the model (7.3.1).

Theorem 7.3.1. (i) $\mathbf{I}_0(\boldsymbol{\tau}) \ge \mathbf{I}(\boldsymbol{\tau})$ and (ii) $\mathbf{I}_0(\boldsymbol{\beta}) \ge \mathbf{I}(\boldsymbol{\beta})$.

Proof. (i) is immediate, (ii) is also immediate since $\mathbf{I}(\boldsymbol{\beta}) = \mathbf{S}_W \le \mathbf{S}_T = \mathbf{I}_0(\boldsymbol{\beta})$.

Corollary 7.3.1. (i) A balanced design minimizes the variance of any treatment contrast estimate. Further, it is **A**–, **D**– and **E**–optimal.
(ii) A balanced design minimizes $V(\hat{\beta}_m)$ for each m.

Remark 7.3.1. It may be noted that for $(\boldsymbol{\tau},\boldsymbol{\beta})$ as a whole, it is *not* true that $\mathbf{I}_0\binom{\boldsymbol{\tau}}{\boldsymbol{\beta}} \ge \mathbf{I}\binom{\boldsymbol{\tau}}{\boldsymbol{\beta}}$.

We shall now make some remarks on **D**–optimality. One might be interested in the estimation of regression coefficients in addition to the estimation of treatment contrasts. We shall show that for **D**–optimality the following inference problems are equivalent in the sense that the same design will be **D**–optimal in each case.

(i) Estimation of $(\mu+\bar{\tau},\mathbf{P}\boldsymbol{\tau},\boldsymbol{\beta})$

(ii) Estimation of $(\mu+\bar{\tau},\mathbf{P}\boldsymbol{\tau})$

(iii) Estimation of $(\mathbf{P}\boldsymbol{\tau},\boldsymbol{\beta})$

(iv) Estimation of $(\mathbf{P}\boldsymbol{\tau})$

where, $\bar{\tau} = \sum_j n_j\tau_j / \sum_j n_j$ and $\mathbf{P}$ is a $(v-1)\times v$ matrix of rank $v-1$ which satisfies $\mathbf{PJ}_{v-1} = \mathbf{0}$.

To see this we consider model (7.3.1) with $\mu+\tau_j = \alpha_j$; $j = 1,2, \ldots, v$. The model can now be written in the matrix form as

$$E(\mathbf{Y}) = [\mathbf{X},\mathbf{Z}]\binom{\alpha}{\beta} = \mathbf{M}\gamma. \tag{7.3.6}$$

We note that

$$|\mathbf{M}'\mathbf{M}| = |\mathbf{Z}'\mathbf{Z}|\ |\mathbf{X}'(\mathbf{I}-\mathbf{P_Z})\mathbf{X}| \tag{7.3.7}$$

where $\mathbf{P_Z}$ is the projection matrix on the column-space of $\mathbf{Z}$. Further, the choice of a design consists in choosing columns of $\mathbf{X}$ in such a way that the j-th column has n_j elements equal to unity and others zero. Moreover, in each row of $\mathbf{X}$ precisely one element is unity. Also, $\text{rank}[\mathbf{X},\mathbf{Z}] = \text{rank}\ \mathbf{M}$ must be $v+k$. As remarked earlier, for each competing design $\mathbf{Z}$ is fixed. We also note that $\mathbf{X}'(\mathbf{I}-\mathbf{P_Z})\mathbf{X}$ is the information matrix for the estimation of α. Since the maximization of $|\mathbf{M}'\mathbf{M}|$ is equivalent to the maximization of $|\mathbf{X}'(\mathbf{I}-\mathbf{P_Z})\mathbf{X}|$, it follows that problems (i) and (ii) are equivalent. This argument is due to Harville (1975).

To establish the equivalence of (i) and (iv) we note that the column space of $\mathbf{X}$ is the same as the space spanned by $\mathbf{1}$ and its orthogonal complement. This orthogonal complement can be used to yield a design matrix for treatment contrast parameters. Since $[\mathbf{Z},\mathbf{1}]$ is again the same for all competing designs, a product decomposition of $|\mathbf{M}'\mathbf{M}|$ similar to (7.3.7) establishes equivalence of problems (i) and (iv). Equivalence of (i) and (iii) is seen by noting that vector $\mathbf{1}$ is the same for all designs.

We have already seen that if a balanced design exists it is $\mathbf{D}$–optimal for problem (iv). It then follows that such a design is also $\mathbf{D}$–optimal for problems (i), (ii) and (iii). We note that $\mathbf{S}_T$ is fixed and for a balanced design $\mathbf{S}_B = 0$. Thus, a balanced design maximizes $\mathbf{S}_W$ in some sense (e.g. $|\mathbf{S}_W|$ is maximum). Of course, in the unbalanced case, i.e. when $\mathbf{S}_B \neq 0$, we do not get strong optimality properties as described in Theorem 7.3.1. However, a design which maximizes an appropriate functional of $\mathbf{S}_W\ (= \mathbf{S}_T - \mathbf{S}_B)$ can be expected to be nearly optimal. The following analysis gives precise results when $k = 1$. In this case we may write down the $\mathbf{C}$–matrix as

$$\mathbf{C} = diag(n_1,n_2,\ldots,n_v) - \begin{pmatrix} n_1 & n_1u_1 \\ \cdot & \cdot \\ \cdot & \cdot \\ \cdot & \cdot \\ n_v & n_vu_v \end{pmatrix} \begin{pmatrix} \dfrac{n_1}{N} & \cdots & \dfrac{n_v}{N} \\ \dfrac{n_1u_1}{T} & \cdots & \dfrac{n_vu_v}{T} \end{pmatrix}$$

where $u_j = \sum_i z_i x_{ij}/n_j - \bar{z}$, $\bar{z} = \sum z_i/N$ and $T = \sum(z_i-\bar{z})^2$. It can be verified that

$$tr\ (\mathbf{C}^+) = \sum_{i=1}^{v} \frac{1}{n_i}(1 - \frac{1}{v}) + \frac{1}{W}[\sum(u_i-\bar{u})^2]$$

where $\bar{u} = \sum u_i / v$ and $W = \sum_i z_i^2 - \sum_j ((\sum_i z_i x_{ij})^2 / n_j)$ is the within treatments sum of squares for z.

Also the characteristic polynomial of $\mathbf{C}$ is $f(\theta) = (\det \mathbf{A})(1 - \mathbf{x}'\mathbf{A}^{-1}\mathbf{x})$ where $\mathbf{A} = ((a_{ij}))$ with $a_{ij} = (n_i - \theta)\delta_{ij} - n_i n_j / N$ and $\mathbf{x}' = (n_1 u_1, \ldots, n_v u_v)$.

We have already seen that when $u_1 = \cdots = u_v = 0$, the design has very strong optimality properties. Also when $n_i = N/v$ for each i, it can be easily seen that each of **A**– and **E**–optimalities requires maximization of W. It was already seen that in this case the design is also **D**–optimal. Thus in the case of equireplicate designs the same designs are **A**–, **D**– and **E**–optimal. When the replication numbers are unequal, this does not hold. This was also observed by Harville (1975) for the estimation of treatment effects (as opposed to treatment contrasts as done here).

For $k>1$, one may devise a measure of lack of balance and may search for a design which minimizes this quantity. A reasonable measure of imbalance is given by

$$\mathbf{IB} = \sum_{m=1}^{k} \lambda_m [\sum_{j=1}^{v} n_j (\bar{z}_{jm} - \bar{z}_{.m})^2]$$

where λ_m is a measure of importance of the m-th covariate.

The above measure is analogous to one introduced by Wu (1981) for the case where the covariates are qualitative. He considered a set up where the m-th covariate takes one of r_m values; $m = 1, 2, \ldots, k$. Let $n_j(u_m)$ denote the number of e.u.'s receiving j-th treatment which have value u_m for the m-th covariate. The measure of imbalance considered by Wu (1981) is given by

$$\mathbf{IB} = \sum_{m=1}^{k} \lambda_m \sum_{u_m=1}^{r_m} \{\sum_{j=1}^{v} n_j^2(u_m) - \frac{1}{v} n^2(u_m)\}$$

where $n(u_m) = \sum_{j=1}^{v} n_j(u_m)$ and λ_m is a weight which reflects the importance of the m-th covariate. A design which is balanced has $\mathbf{IB} = 0$. When such a design does not exist a design which minimizes **IB** can be expected to be a good design. A design which minimizes **IB** is called **B**–optimal. A sufficient condition for **B**–optimality is that $|n_j(u_m) - \frac{1}{v} n(u_m)| < 1$ for all j and all m.

Wu termed designs with this property as designs with *maximal balance*. Clearly, a design with maximal balance is **B**–optimal for any choice of the λ_m's.

Since the search for **B**–optimal designs may be very expensive, Wu (1981) recommends a search for designs which are nearly **B**–optimal and gives an algorithm for finding such designs. Finally, he gives some examples of the application of this algorithm. In one of these examples the algorithm gives a design with maximum balance. Extensions of **B**–optimality such as inclusion of interactions among covariates are discussed by Wu (1981).

An important step in Wu's algorithm is the Switch-Exchange routine. The switch part of this routine involves switching an e.u. from one treatment to another and as such affects treatment replication numbers. Thus, as noted by Wu, with the use of this algorithm, the experimenter has no control over replication numbers but it seems reasonable to expect that the replication numbers will not be far from approximate equality.

It should also be noted that the notion of **B**–optimality is *not directly* related to statistically meaningful criteria such as **A**–, **D**– or **E**–optimality. However, the results of Haggstrom indicate that **B**–optimal designs will perform well w.r.t. these criteria.

It may be remarked that the model considered by Wu (1981) is in fact the model where one compares v treatments in the presence of arbitrary k–dimensional blocking. If $k = 1$ and block sizes are equal or if $k = 2$ and the blocking in two dimensions is of row-column type, results of Chapters Two, Three or Four will be applicable.

Lopes Troya (1982a, 1982b) considered the case where the values of the covariates are not fixed but are to be chosen by the experimenter. She considers the problem of constructing **D**–optimal designs for the joint estimation of the parameter (β,τ). Since τ can be replaced by the general mean and a set of contrasts and since the coefficient of the general mean is the same for any e.u., **D**–optimality of (β,τ) is *equivalent* to **D**–optimality for the joint estimation of β and any linearly independent set of $(v-1)$ treatment contrasts.

If the z_{im}'s be constrained to take values in $[-1,1]$ Lopes Troya (1982a) has shown that for a **D**–optimal design (for the estimation of all parameters), every z_{im} must be ± 1. She also gave the following table which gives the maximum possible value for the determinant of the information matrix for all the $v+k$ parameters for an equireplicate design. In this table n denotes the common replication number.

Table 7.3.1: Maximum value for $|\mathbf{M'M}|$

(a)	$N \equiv 0(mod\ 4)$, n even	$n^v \times N^k$
(b)	n odd	$n^v \times (N-v/n)^k$
(c)	$N \equiv 2(mod\ 4)$, $n = 2$	$n^v \times (N-2)^{k-1}(N-2-2k)$
(d)	$N \equiv 2(mod\ 4)$, $n \equiv 2(mod\ 4)$, $n > 2$ and $k \leq (N-2)/(n-2)$	$n^v \times (N-2)^{k-1}(N-2-2k)$
(e)	$N \equiv 2(mod\ 4)$, $n \equiv 2(mod\ 4)$, $n>2$ and $k>(N-2)/(n-2)$	$n^v \times (N-2)^{k-2}(N-2+k+\frac{N-2}{n-2})(N-2+k-\frac{2k}{n}-\frac{N-2}{n})$

Of course, designs for which these bounds are attained need not exist in all cases. Methods of construction of designs attaining the above bounds for many cases are given in Lopes Troya (1982a). It is clear that for a design with fixed values of $n_1,n_2, \ldots, n_v$, **D**–optimality for the estimation of the regression coefficients is equivalent to **D**–optimality for the estimation of all the effects.

However, even when $n_1, n_2, \ldots, n_v$ are fixed, a design which is **D**–optimal for all effects need not be **D**–optimal for the estimation of treatment contrasts. An example of this is provided by the following design due to Lopes Troya (1982a). For this design, $v = 5$, $n_1 = n_2 = n_3 = n_4 = n_5 = 3$ and $k = 2$. For each of the five treatments we give below the values of the two covariates z_1 and z_2.

Treatment	Values of (z_1,z_2) for the three e.u.'s		
1	(-1,1)	(1,-1)	(-1,1)
2	(1,1)	(-1,-1)	(1,-1)
3	(1,1)	(-1,-1)	(-1,1)
4	(-1,-1)	(1,1)	(1,-1)
5	(1,-1)	(-1,1)	(1,1)

This design is **D**–optimal for the joint estimation of the treatment contrasts and the regression coefficients when the n_j's are fixed. However, for each covariate, the means of the covariate values for different treatments are unequal and hence the design is *not* **D**–optimal just for the estimation of treatment contrasts. We note that this follows from the fact that one can choose covariate values in the range [–1,1] in such a way that for each covariate the mean of covariate values is the same for each treatment and such a design is **D**–optimal for the estimation of treatment contrasts.

4. Designs for Comparing Treatments vs. Control

So far we have discussed various results on the optimality aspects of block designs and row-column designs with reference to optimality functionals of the type $\Phi(\mathbf{C})$ with Φ satisfying certain properties. Quite naturally, these results are built upon the supposition that the experimenter has *equal* interest in all treatment comparisons. This, however, should *not* give the impression that the inference problems are always *invariant* w.r.t. permutations of the treatments under consideration. While this could be a general phenomenon with many experiments, an experimenter, on a given occasion, might be interested in some specific treatment comparisons which may *not* necessarily remain unchanged w.r.t. permutations of *all* the treatments. A common example would be that of comparing a set of test treatments with one or more control treatments. We will primarily ellaborate on this problem in this section. In recent years, optimality studies related to such test vs. control treatment comparisons have been drawing increasing attention of researchers in this area.

Kiefer (1958) derived the optimality properties of the **BBDs** and the **GYDs** with reference to the problem of inference on a full set of orthonormal treatment contrasts. Writing $\begin{pmatrix} \mathbf{1}'/\sqrt{v} \\ \mathbf{P} \end{pmatrix}$ for an orthogonal matrix and τ for the vector of treatment effects parameters, the inference problem involved the contrasts $\mathbf{P}\tau$ for which the best linear unbiased estimate (**BLUE**) is $\mathbf{P}\hat{\tau}$ with the variance-covariance matrix (under the usual fixed-effects block-design set-up, for example) given by $\sigma^2(\mathbf{PC}^+\mathbf{P}')$. Clearly, only connected designs are of relevance here. Using the algebraic properties of the **P**-matrix viz., $\mathbf{PP}' = \mathbf{I}_{v-1}$ and $\mathbf{P}'\mathbf{P} = \mathbf{I}_v - \mathbf{J}_v/v$, it can be shown that

$(\mathbf{PC^+P'}) = (\mathbf{PCP'})^{-1}$. Moreover, the eigenvalues of $\mathbf{PCP'}$ are the positive eigenvalues $x_1, x_2, \ldots, x_{v-1}$ of the $\mathbf{C}$–matrix. This means that the eigenvalues of the variance-covariance matrix of $\mathbf{P}\hat{\tau}$ are given by (apart from the multiplier σ^2) x_1^{-1}, $x_2^{-1}, \ldots, x_{v-1}^{-1}$. Hence, w.r.t. the usual optimality criteria like $\mathbf{A}$–, $\mathbf{D}$– or $\mathbf{E}$–optimality, the optimality problems amount to minimizing $\sum \frac{1}{x_i}$, Πx_i^{-1} or $\max_i(x_i^{-1})$. This directly leads to the optimality of the **BBDs** w.r.t. the above three criteria.

What happens if the inference problem is different from the above? At the very outset, we should make it clear that for *arbitrary* inference problems involving treatment comparisons, the search for an optimal design is indeed a very difficult task. However, sometimes, there may exist *some symmetry* in the inference problem. If properly exploited, this may lead to a simplified approach in the search for an optimal design. This motivated Sinha (1970, 1972a, 1972b, 1975, 1980, 1982) to study in detail various structures of inference problems and exploit them in a proper manner. The specific study of test vs. control treatment comparisons also rests on this approach to a large extent. This study was initiated in Bechhofer and Tamhane (1981) and resolved satisfactorily by others in a series of papers. Following Sinha (1970, 1972a, 1972b, 1982), we will attempt a general discussion of this approach below. Then we will take up the particular problem of test vs. control treatments comparison and present the solution due to Mazumdar and Notz (1983).

We start with an inference problem, to be denoted by π, involving some treatment comparisons.

$$\pi: \ \boldsymbol{\eta}(i\times 1) = \mathbf{L}(i\times v)\boldsymbol{\tau}(v\times 1), \quad \mathbf{L1} = \mathbf{0} \tag{7.4.1}$$

We will say that π is a *full-rank non-singular* problem iff rank $(\mathbf{L}) = i = v-1$. It will be said to be *non-singular* iff rank $(\mathbf{L}) = i$, and of *full rank* iff $\text{rank}(\mathbf{L}) = v-1$. Only full rank problems have so far been largely dealt with in the literature. Note that full-rank problems call for a study of only the connected designs in a given set-up.

As usual, we denote by $\mathbf{S}_v$ the symmetric group of $v!$ permutations of the treatment effects $\tau_1, \tau_2, \ldots, \tau_v$. Also $g, h, \ldots$ will be taken as members of $\mathbf{S}_v$ and the corresponding permutation matrices will be denoted by $\mathbf{G}_g, \mathbf{G}_h$ etc. Denote by $\bar{\mathbf{S}}_i$ the symmetric group of $i!$ permutations of the components $\eta_1, \eta_2, \ldots, \eta_i$ of $\boldsymbol{\eta}$ in (7.4.1). Members of $\bar{\mathbf{S}}_i$ will be denoted by the symbols $\bar{g}, \bar{h}, \ldots$ with the corresponding permutation matrices written as $\mathbf{G}_{\bar{g}}, \mathbf{G}_{\bar{h}}$ etc.

We start with the following definitions.

Definitions. The problem π is defined to be invariant w.r.t. the permutation $g \epsilon \mathbf{S}_v$ if there is a permutation $\bar{g} \epsilon \bar{\mathbf{S}}_i$ such that $\mathbf{G}'_{\bar{g}}\mathbf{L}\mathbf{G}_g = \mathbf{L}$. Moreover, π is said to be invariant w.r.t. a subgroup of permutations if it is so w.r.t. every permutation of the subgroup. When the subgroup coincides with $\mathbf{S}_v$, we call π symmetrically invariant. An interesting subgroup is $\mathbf{S}_{v-1}(1)$ which is the subgroup of all $(v-1)!$ permutations leaving τ_1 unchanged.

Suppose the problem π is invariant w.r.t. the permutation $g \epsilon \mathbf{S}_v$. Then changing τ to $\mathbf{G}_g'\tau = \tau_g$ (say), the problem π changes to π_g: $\eta_g = \mathbf{L}\tau_g = \mathbf{L}\mathbf{G}_g'\tau = \mathbf{G}_{\bar{g}}'\mathbf{L}\tau = \mathbf{G}_{\bar{g}}'\eta$. This shows that for a g–invariant problem π, the derived problem π_g is related to the problem π through a permutation $\mathbf{G}_{\bar{g}}$. We refer to Sinha (1970, 1972a, 1972b, 1975) for various results on characterizations of appropriately invariant problems w.r.t the symmetric group or its subgroup of permutations. It may be noted that the problem of inference on *all* elementary treatment contrasts in a symmetrically invariant *singular* full-rank problem.

From now onwards, we will take π to refer to a *non-singular* full-rank problem i.e., π: $\eta((v-1)\times 1) = \mathbf{L}((v-1)\times v)\tau(v\times 1)$, $\mathbf{L1} = \mathbf{0}$, $\text{rank}(\mathbf{L}) = v-1$. Suppose π is invariant w.r.t. a subgroup $\mathbf{G}$ of $\mathbf{S}_v$. Thus, for every member $g \epsilon \mathbf{G}$, there is a corresponding member $\bar{g}$ (say) $\epsilon\ \bar{\mathbf{S}}_{v-1}$ such that $\mathbf{G}_{\bar{g}}'\mathbf{L}\mathbf{G}_g = \mathbf{L}$ and, hence, $\mathbf{G}_{\bar{g}}\mathbf{L}\mathbf{G}_g' = \mathbf{L}$ as well. Consider now a block design set-up with b blocks involving k experimental units each and with the v treatments under comparison being distributed over the bk experimental units. We assume the usual fixed-effects additive model to hold. Denote by $\mathbf{C}$ the $\mathbf{C}$–matrix of a design used for drawing inference on the statistical problem π. As π is of full-rank, only connected designs are of relevance here. The BLUE of η is given by $\hat{\eta} = \mathbf{L}\hat{\tau}$ with the variance-covariance matrix $\sigma^2(\mathbf{L}\mathbf{C}^+\mathbf{L}')$. In other words, the information matrix for $\hat{\eta}$ is given by

$$\mathbf{I}(\hat{\eta}) = \sigma^{-2}(\mathbf{L}\mathbf{C}^+\mathbf{L}')^{-1}. \tag{7.4.2}$$

Our object is to unify the search for an optimal design in the class of relevant designs for a general optimality functional Φ defined on the class of information matrices $\mathbf{I}(\hat{\eta})$. We will assume Φ to be convex and symmetric but otherwise, keep it arbitrary. The main result of Sinha (1980, 1982) is that corresponding to any $\mathbf{C}$–matrix, there exists a $\mathbf{G}$-invariant $\mathbf{C}$–matrix, say, $\tilde{\mathbf{C}}$ such that $\tilde{\mathbf{C}}$ performs better than $\mathbf{C}$ in the sense that

$$\Phi(\tilde{\mathbf{I}}(\hat{\eta})) = \Phi(\sigma^{-2}(\mathbf{L}\tilde{\mathbf{C}}^+\mathbf{L}')^{-1}) \leq \Phi(\sigma^{-2}(\mathbf{L}\mathbf{C}^+\mathbf{L}')^{-1}) = \Phi(\mathbf{I}(\hat{\eta})). \tag{7.4.3}$$

Here $\tilde{\mathbf{I}}$ corresponds to the information matrix based on $\tilde{\mathbf{C}}$. Further, $\mathbf{G}$–invariance of $\tilde{\mathbf{C}}$ is to be interpreted as the matrix $\tilde{\mathbf{C}}$ satisfying the property $\mathbf{G}_g'\tilde{\mathbf{C}}\mathbf{G}_g = \tilde{\mathbf{C}}$ for all $g \epsilon \mathbf{G}$.

The above inequality has an interesting and appealing consequence. This is that a *formal* search for the optimal design might be confined to a search for an optimal $\mathbf{G}$–*invariant* $\mathbf{C}$–matrix.

Since such $\mathbf{G}$–invariant $\mathbf{C}$–matrices will have nice algebraic structures, the search for one minimizing an optimality functional $\Phi(.)$ may be relatively easier. Of course, the final task would be to *relate* such an optimal $\mathbf{C}$–matrix to a *concrete* design in the given experimental set-up. This last task is certainly *not* at all easy all the time. We will discuss more about it later.

We now give the form of such a $\tilde{\mathbf{C}}$–matrix corresponding to a given $\mathbf{C}$–matrix.

This is derived from the following relations (Vide Sinha (1982)).

(i) $(\mathbf{L\tilde{C}^{+}L'})^{-1} = \sum_{g\epsilon \mathbf{G}}\{\mathbf{L}(\mathbf{G}_g'\mathbf{C}^{+}\mathbf{G}_g)\mathbf{L}'\}^{-1}/\#(\mathbf{G}),$ (7.4.4)

(ii) $\tilde{\mathbf{C}}$ has rank $(v-1)$ and $\tilde{\mathbf{C}}\mathbf{1} = \mathbf{0}$, (7.4.5)

Here $\#(\mathbf{G})$ denotes the cardinality of the subgroup $\mathbf{G}$.

We refer to Sinha (1982) for a proof of the facts that such a $\tilde{\mathbf{C}}$ is unique and that $\tilde{\mathbf{C}}$ is $\mathbf{G}$–invariant. For the rest of this section, we will specialize to the problem of test vs. control treatment comparisons and present known optimality results, due mainly to Bechhofer and Tamhane (1981) and Majumdar and Notz (1983).

A slight change of the basic set-up and notations is desired at this stage. We will assume that there are v test treatments and one control treatment so that the total number of treatments is $(v+1)$. We will denote by $\tau_1, \ldots, \tau_v$ the effects of the test treatments and by τ_0 that of the control treatment. The specific problem of interest to us is to infer about

$$\pi:\ \boldsymbol{\eta}(v\times 1) = (\tau_0-\tau_1, \tau_0-\tau_2, \ldots, \tau_0-\tau_v)'. \tag{7.4.6}$$

Clearly, π is invariant w.r.t. the subgroup $\mathbf{S}_v(0)$ (of $\mathbf{S}_{v+1}$) which keeps the control treatment (denoted by 0) intact and permutes all the test treatments among themselves. Accordingly, given any $\mathbf{C}$–matrix, there exists an $\mathbf{S}_v(0)$–invariant $\tilde{\mathbf{C}}$–matrix which performs better w.r.t. any convex symmetric criterion $\Phi(.)$.

Clearly, the $\tilde{\mathbf{C}}$–matrix is characterized by the conditions (i) and (ii) displayed above where, for the above choice of π, we have

$$L = [\mathbf{1}|(-1)\mathbf{I}_v] \tag{7.4.7}$$

and, accordingly, as noted above, $\mathbf{G} = \mathbf{S}_v(0)$ with $\#(\mathbf{G}) = v!$.

Bechhofer and Tamhane (1981) have made a very interesting and pertinent observation at this stage. Writing $\mathbf{C} = ((c_{ij}))$, $i,j = 0,1,2, \ldots, v$, they have shown that, for the above choice of $\mathbf{L}$,

$$(\mathbf{LC^{+}L'})^{-1} = \mathbf{C}_{00} = ((c_{ij}))\quad i,j = 1,2, \ldots, v \tag{7.4.8}$$

where $\mathbf{C}_{00}$ is used to denote the submatrix of $\mathbf{C}$ deleting the 0-th row and column.

When $\mathbf{G} = \mathbf{S}_v(0)$ and $\text{rank}(\mathbf{L}) = v$, it has been shown in Sinha (1970) that corresponding to g, $h\epsilon\mathbf{G}$, $g \neq h$, the permutations $\bar{g}$ and $\bar{h}$ are also different. Here $\mathbf{G}_{\bar{g}}'\mathbf{L}\mathbf{G}_g = \mathbf{L} = \mathbf{G}_{\bar{h}}'\mathbf{L}\mathbf{G}_h$. Note, incidentally, that there are precisely $v!$ permutations of the type $\bar{g}$ to match with the same number of permutations in $\mathbf{G} = \mathbf{S}_v(0)$.

We may now rewrite (7.4.4), using (7.4.5), as

$$(L\tilde{C}^{+}L')^{-1} = \tilde{\mathbf{C}}_{00} = \sum_{\bar{g}\epsilon\bar{\mathbf{S}}_v} \{\mathbf{G}_{\bar{g}}'(\mathbf{LC^{+}L'})\mathbf{G}_{\bar{g}}\}^{-1}/v!$$

$$= \sum_{\bar{g}\epsilon\bar{\mathbf{S}}_v} (\mathbf{G}_{\bar{g}}'\mathbf{C}_{00}\mathbf{G}_{\bar{g}})/v!$$

$$= (a-b)\mathbf{I}_v + b\mathbf{J}_{v\times v} \tag{7.4.9}$$

where $a = \sum_1^v C_{ii}/v$ and $b = \sum\sum_{1\leq i<j\leq v} C_{ij}/\binom{v}{2}$. This shows that the matrix $\tilde{\mathbf{C}}$ is necessarily of the form*

$$\tilde{\mathbf{C}} = \begin{bmatrix} \alpha & \beta & \beta & \cdots & \beta \\ \beta & a & b & \cdots & b \\ \beta & b & a & \cdots & b \\ \cdot & \cdot & \cdot & \cdot & \vdots \\ \beta & b & \cdot & \cdots & a \end{bmatrix} \tag{7.4.10}$$

Majumdar and Notz (1983) have directly shown that the above $\tilde{\mathbf{C}}$–matrix performs better than the $\mathbf{C}$–matrix we start with for *any* convex symmetric criterion $\Phi(.)$ for the inference problem π in (7.4.6).

The problem thus comes down to comparing various $\tilde{\mathbf{C}}$–matrices with the formal explicit forms as specified above in (7.4.10). The optimality function Φ is taken as a convex symmetric real-valued function on the class of all p.d. matrices which naturally includes all matrices of the form $\mathbf{C}_{00}$.

For a given connected design with design parameters $(b,v+1,k)$ satisfying $v+1<k$, let $\mathbf{N}$ denote the usual incidence matrix so that $\mathbf{N} = ((n_{ij}))$, $i = 0,1,\ldots,v$, $j = 1,2,\ldots,b$. Write $r_0 = \sum_j n_j$, $r_i = \sum_j n_{ij}$, $1\leq i\leq v$. Let $\lambda_{ii'} = \sum_j n_{ij}n_{i'j}$, $i,i' = 0,1,\ldots,v$. It follows that $\tilde{\mathbf{C}}_{00} = (a-b)\mathbf{I}_v + b\mathbf{J}_{v\times v}$ where

$$a = \{\sum_1^v r_i - \sum_{i=1}^v \lambda_{ii}/k\}/v \qquad b = -\sum\sum_{1\leq i\neq i'\leq v} \lambda_{ii'}/kv(v-1). \tag{7.4.11}$$

Further, from (7.4.10),

---

*The combinatorial and constructional aspects of designs having $\mathbf{C}$–matrices of the form (7.4.10) or of forms more general than the one occurring in this context have been thoroughly investigated by several researchers. See, for example, Adhikary (1965). Rao (1947) calls such designs *Intra-* and *Inter-Group Balanced Block Designs* (**IIGBBDs**). The reinforced **BIBDs** introduced by Das (1958) also have such forms of $\mathbf{C}$–matrices. Bechhofer and Tamhane (1981) called such designs as *Balanced Treatment Incomplete Block* (**BTIB**) designs.

$$a + (v-1) = -\beta = \frac{\alpha}{v} = \{r_0 - \frac{\sum n_{0j}^2}{k}\}/v \tag{7.4.12}$$

and

$$a-b = [tr(\bar{\mathbf{C}}_{00}) - \{a+(v-1)b\}]/(v-1)$$

$$= [(\sum r_i - \frac{\sum \lambda_{ii}}{k}) - (r_0 - \frac{\sum r_{0j}^2}{k})/v]/(v-1). \tag{7.4.13}$$

The first step in getting optimal designs is to rule out the non-binary designs, specifically the ones which are non-binary in the test treatments. This is accomplished in the following lemma.

Lemma 7.4.1. Suppose a design with incidence matrix $\mathbf{N}$ is *not* binary in the test treatments. Then we can construct a design with incidence matrix $\mathbf{N}^*$ which is binary in the test treatments and also satisfies $\Phi(\bar{\mathbf{C}}_{00}) \geq \Phi(\bar{\mathbf{C}}_{00}^*)$ provided that Φ is such that $\Phi(\mathbf{A}) \geq \Phi(\mathbf{A}^*)$ for two p.d. matrices $\mathbf{A}$ and $\mathbf{A}^*$ which are of the form

$$\mathbf{A} = (a-b)\mathbf{I}_v + b\mathbf{J}_{v\times v},\ A^* = (a^*-b^*)\mathbf{I}_v + b^*\mathbf{J}_{v\times v}$$

with

$$a^* + (v-1)b^* \geq a + (v-1)b \text{ and } a^* - b^* \geq a-b.$$

Proof. The proof is by construction of $\mathbf{N}^*$ from given $\mathbf{N}$. In each block of the design we start with, we replace any duplicates of test treatments by other test treatments *not* in the block so that in effect the modified design with incidence matrix $\mathbf{N}^*$ is *binary* in the test treatments. Note that in the process, n_{0j}'s and, hence, r_0 remain unchanged. Further, $\sum r_i = \sum r_i^*$ and since $\mathbf{N}^*$ is binary, we also have $\sum \lambda_{ii} = \sum_i \sum_j n_{ij}^2 \geq \sum\sum n_{ij} = \sum r_i = \sum r_i^* = \sum\sum n_{ij}^* = \sum_i \sum_j n_{ij}^{*2} = \sum \lambda_{ii}^*$. We now identify $\bar{\mathbf{C}}_{00}$ as $\mathbf{A}$ and $\bar{\mathbf{C}}_{00}^*$ as $\mathbf{A}^*$ in our statement of the Lemma. Referring to (7.4.12) and (7.4.13), we readily see that $a+(v-1)b = a^*+(v-1)b^*$ and $a-b \leq a^*-b^*$ so that $\mathbf{A} \leq \mathbf{A}^*$. Thus $\Phi(\mathbf{A}) \geq \Phi(\mathbf{A}^*)$ which is the same as $\Phi(\bar{\mathbf{C}}_{00}^*) \leq \Phi(\bar{\mathbf{C}}_{00})$. This proves the lemma.

The next result helps to eliminate all designs with $r_0 > \frac{bk}{2}$.

Lemma 7.4.2. Suppose we start with a design having incidence matrix $\mathbf{N}$ and it is binary in the test treatments and yields $r_0 > \frac{bk}{2}$. Then we can construct a design with incidence matrix $\mathbf{N}^*$ which is also binary in test treatments, has $r_0 < bk/2$ and satisfies $\Phi(\bar{\mathbf{C}}_{00}^*) \leq \Phi(\bar{\mathbf{C}}_{00})$ for any optimality function Φ defined as in Lemma 7.4.1.

Proof. The proof is again by construction of $\mathbf{N}^*$ from given $\mathbf{N}$. For every j, $1 \leq j \leq b$, we take $n_{0j}^* = \sum_{i=1}^{v} n_{ij}$ and $n_{ij}^* = 0$ if $n_{ij} = 1$. If $n_{ij} = 0$, choose $n_{ij}^* = 0$ or 1 subject to $\sum_{1}^{v} n_{ij}^{*} = n_{0j}$. Then $r_0^* = bk-r_0<\frac{bk}{2}<r_0$ and $r_0^*-\sum_{j=1}^{b} n_{0j}^{*2}/k = r_0-\sum_{j=1}^{b} n_{0j}^2/k$. The rest of the proof follows along the line of Lemma 7.4.1.

We now specialize to functional forms of Φ of the type $\Phi(\mathbf{C}_{00}) = \sum_{1}^{v} f(\mu_i)$ where $\mu_1, \ldots, \mu_v$ are the eigenvalues of $\mathbf{C}_{00}$ and f is a real-valued function on the set of all positive real numbers, is continuously differentiable and has $f'<0$ and $f''>0$.

We state the following result (without proof) from Majumdar and Notz (1983). Below $[x]$ represents the integral part of x.

Theorem 7.4.1. Suppose there exists a design with incidence matrix $\mathbf{N}^*$ such that $\mathbf{C}_{00}^*$ is completely symmetric and

(i) the design is binary in test treatments,

(ii) the control treatment replication number r_0^* minimizes

$$g(r,b,v+1,k) = f(\{r-h(r,b)/k\}/v) + (v-1)f(\{b(k-1)-\frac{k-1}{k}r-(r-h(r,b,)/k)/v\}/(v-1))$$

where

$$h(r,b) = \{b(1+[r/b])-r\}[r/b]^2+(r-b[r/b])([r/b]+1)^2$$

(iii) n_{0j}^*'s assume values $[r_0^*/b]$ or $[r_0^*/b]+1$ for all $1\leq j\leq b$.

Then the design is Φ–optimal among all connected designs for an optimality functional Φ of the form indicated above.

Two particular choices of f would be $f(x) = -\log x$(**D**–optimality) and $f(x) = \frac{1}{x}$(**A**–optimality). For **A**–optimality, the function $g(.)$ in (ii) above assumes the form

$$g(r,b,v+1,k) = v/\{r - \frac{h(r,b)}{k}\} + (v-1)^2/\{b(k-1) - \frac{r(k-1)}{k} - v^{-1}(r - \frac{h(r,b)}{k})\} \quad (7.4.14)$$

For **D**–optimality, the problem has a ready-made solution from a different consideration. Recall the invariance property of **D**–optimality criterion mentioned in Remark 1.3.2. It then follows that a **D**–optimal design is *independent* of the *specific* form of the non-singular full-rank problem under consideration. In other words, we can outright start with $\eta = \mathbf{P}\tau$ for the **D**–optimality criterion. Then the result is that a **BBD**, whenever it exists, is **D**–optimal for *any* non-singular full-rank problem. Again, as regards **E**–optimality, altogether different considerations are needed. We will *not* pursue **E**–optimality problems here. We refer to Hedayat, Jacroux and Majumdar (1988) for more recent results on this topic.

Theorem 7.4.1 has been exploited exhaustively and various results have been made available on **A**–optimal designs for test treatments vs. control comparisons in a series of papers by Hedayat and Majumdar (1983, 1984, 1985). See also Ture (1982), and Notz and Tamhane (1983). More recent results along this line of research are available in Hedayat and Majumdar (1986) and Stufken (1987). A catalogue of such A–optimal designs for $2 \le k \le 8$, $k \le v \le 30$, $v \le b \le 50$ has also been made available in Hedayat and Majumdar (1984). Sinha (1980, 1982) demonstrated that in the restricted class of *equireplicate* designs for given b, $v+1$ and k, a **BIBD**, if it exists, is still $\Phi(.)$–optimal for *any* convex optimality functional $\Phi(.)$. On the other hand, in the unrestricted class of *all* connected designs, the efficiency factor of a **BIBD** w.r.t. the hypothetical A–optimum design is given by

$$\left. \begin{aligned} E &= \frac{(v-3)^2}{2(v-1-\sqrt{v+1})^2} && \text{if } v \neq 3 \\ &= \frac{8}{9} && \text{if } v = 3 \end{aligned} \right\} \qquad (7.4.15)$$

See also Hedayat and Majumdar (1984) in this context.

It follows that the limiting efficiency of a **BIBD** is only 50% for control treatment vs. test treatment comparisons.

We now turn back to Theorem 7.4.1 and examine the steps to be followed in determining specific A–optimal designs. Referring to the expression (7.1.14), for given b,v and k, first we have to determine r_0^* which minimizes $g(r,b,v+1,k)$. Then we have to get hold of a design which is binary in test treatments and yields (i) $n_{0j}^* = [r_0/b]$ or $[r_0/b]+1$ for all $1 \le j \le b$, (ii) r_j's all equal, $\lambda_{jj'}$'s all equal, (iii) λ_{0j}'s all equal. These will ensure complete symmetry of the underlying $\mathbf{C}_{00}$ matrix. We illustrate it with the following example. Consider $b = 24$, $k = 3$, $v = 9$. Majumdar and Notz (1983) have verified that $r_0^* = 18$ minimizes (7.1.14). The following design then satisfies the conditions of Theorem 7.4.1 and is, therefore, A–optimal.

Blocks	Treatments			Blocks	Treatments		
1	0	1	3	13	0	4	9
2	0	1	4	14	0	5	6
3	0	1	5	15	0	6	8
4	0	1	8	16	0	6	9
5	0	2	4	17	0	7	8
6	0	2	5	18	0	7	9
7	0	2	7	19	1	2	9
8	0	2	8	20	1	6	7
9	0	3	5	21	2	3	6
10	0	3	7	22	3	4	8
11	0	3	9	23	4	5	7
12	0	4	6	24	5	8	9

We conclude this section with the observation that determination of r_0^* minimizing (7.1.14) may *not* always be an easy task. Of course, a computer search should enable one to tackle this problem in a convenient manner. Even if this problem is resolved, existence of a design satisfying the other conditions of Theorem 7.4.1 is *not* automatically ensured and there are examples of situations where no such design exists. This really complicates the search for an A–optimal design even in the simplest block design set-up.

Recently, an analogous study of MV-optimal block designs has been made by Jacroux (1986). Also extensions of this problem to row-column designs, to repeated measurements designs and/or to situations involving more than one control treatments has been made (Notz (1985), Majumdar (1986) and Pigeon and Raghavarao (1987).) Further work along this direction is in progress.

REFERENCES

Adhikary, B. (1965). On the properties and construction of balanced block designs with variable replications. *Cal. Statist. Assoc. Bull, 14*, 36-64.

Bechhofer, R.E. and Tamhane, A.C. (1981). Incomplete block designs for comparing treatments with a control: General theory. *Technometrics, 23*, 45-57.

Cheng, C.S. (1983). Construction of optimal balanced incomplete block designs for correlated observations. *Ann. Statist., 11*, 240-246.

Das, M.N. (1958). On reinforced incomplete block designs. *J. Ind. Soc. Agr. Statist. 10*, 73-77.

Gill, P.S. and Shukla, G.K. (1985). Efficiency of nearest neighbour balanced block designs for correlated observations. *Biometrika, 72*, 539-544.

Haggstrom, G.W. (1975). The pitfalls of manpower experimentation. RAND Corporation, Santa Monica, Calif.

Hanani, H. (1961). The existence and construction of balanced incomplete block designs. *Ann. Math. Statist. 32*, 361-386.

Harville, D.A. (1975). Computing optimum designs for covariate models. In a Survey of Statistical Design and Linear Models. (J.N. Srivastava, ed.) Amsterdam: North-Holland Publishing Co., (1975), 209-228.

Hedayat, A.S., Jacroux, M. and Majumdar, D. (1988). Optimum designs for comparing test treatments with controls (with discussion). *Statistical Science, 3*, 462-491.

Hedayat, A.S. and Majumdar, D. (1983). Families of $A-$ optimal block designs for comparing test treatments with a control. *Bull. Inst. Math. Statist., 12*, 278 (Abstract No. 83t-84).

Hedayat, A.S. and Majumdar, D. (1984). $A-$ optimal incomplete block designs for control-test treatment comparisons. *Technometrics, 26*, 363-370.

Hedayat, A.S. and Majumdar, D. (1985). Families of $A-$ optimal block designs for comparing test treatments with a control. *Ann. Statist., 13*, 757-767.

Hedayat, A.S. and Majumdar, D. (1986). Recent discoveries on $A-$ optimal designs for comparing test treatments with controls. Technical Report No. 04-86, Dept. Math., Stat. & Comp. Sc., Univ. Illinois at Chicago, Chicago, Ill.

Jacroux, M. (1986). On the determination and construction of MV-optimal block designs for comparing test treatments with a standard treatment. *J. Statist. Planning and Inference, 15*, 205-226.

Kiefer, J. (1958). On the nonrandomized optimality and randomized nonoptimality of symmetric designs. *Ann. Math. Statist., 29*, 675-699.

Kiefer, J. and Wynn, H.P. (1981). Optimum balanced block and Latin square designs for correlated observations. *Ann. Statist., 9*, 737-757.

Lopes Troya, J. (1982a). Optimal designs for covariate models. *J. Statist. Planning and Inference, 6*, 373-419.

Lopes Troya, J. (1982b). Cyclic designs for a covariate model. *J. Statist. Planning and Inference, 7*, 49-75.

Majumdar, D. and Notz, W. (1983). Optimal incomplete block designs for comparing treatments with a control. *Ann. Stat., 11*, 258-266.

Majumdar, D. (1986). Optimal designs for comparisons between two sets of treatments. *J. Statist. Planning and Inference, 14*, 359-372.

Notz, W. and Tamhane, A.C. (1983). Incomplete block (BTIB) designs for comparing treatments with a control: minimal complete sets of generator designs for $k = 3$, $p = 3(1)10$. *Comm. Statist. (A) Theo. Meth. 12*, 1391-1412.

Notz, W. (1985). Optimal designs for treatment-control comparisons in the presence of two-way heterogeneity. *J. Statist. Planning and Inference, 12*, 61-73.

Pigeon, J.G. and Raghavarao, D. (1987). Cross-over designs for comparing treatments with a control. *Biometrika, 74*, 321-328.

Rao, C.R. (1947). General methods of analysis for incomplete block designs. *Jour. Amer. Statist. Assoc., 42*, 541-561.

Russell, K.G. and Eccleston, J.A. (1987). The construction of optimal balanced incomplete block designs when adjacent observations are correlated. *Austral. J. Statist., 29*, 84-90.

Sinha, Bikas K. (1970). Invariant problems of linear inference and related designs. *Cal. Statist. Assoc. Bull., 19*, 103-122.

Sinha, Bikas K. (1972a). Contribution to comparison of experiments: optimum experiments for linear inference. Unpublished Ph.D. Thesis, Calcutta University.

Sinha, Bikas K. (1972b). Invariant problems of linear inference and related designs

- an addendum. *Cal. Statist. Assoc. Bull.*, *21*, 51-55.

Sinha, Bikas K. (1975). Regular cyclically invariant problems of linear inference and related designs. *Sankhyā (B)*, *37*, 79-90.

Sinha, Bikas K. (1980). Optimal block designs. Unpublished Seminar Notes. Indian Statistical Institute, Calcutta.

Sinha, Bikas K. (1982). On complete classes of experiments for certain invariant problems of linear inference. *J. Statist. Planning and Inference*, *7*, 171-180.

Stufken, J. (1987). *A*-optimal block designs for comparing test treatments with a control. *Ann. Statist.*, *15*, 1629-1638.

Ture, T.E. (1982). On the construction and optimality of balanced treatment incomplete block designs. Ph.D. Thesis, University of California, Berkeley.

Wierich, W. (1985). Optimum designs under experimental constraints for a covariate model and an intra-class regression model. *J. Statist. Planning and Inference*, *12*, 27-40.

Wu, C.F.J. (1981). Iterative construction of nearly balanced assignments I: Categorical covariates. *Technometrics*, *23*, 37-44.

CHAPTER EIGHT

WEIGHING DESIGNS

1. Introduction

So far we have presented various theoretical aspects of optimality studies in the set-up of *traditional* experimental designs. A completely different field of study where optimality considerations have experienced surprisingly distinctive growth in its own merit is the study of *weighing problem* which originated in a casual illustration furnished by Yates (1935). The precise formulation of such problems is to be found in Hotelling (1944). Over the past forty years, various aspects of such problems have been so extensively studied that this topic has already attained the status of *weighing designs* in design of experiments. Raghavarao (1971) has provided a fairly complete account of the basic results available in this area. Banerjee (1975) has introduced the subject matter in general terms to research workers in applied sciences.

A study of weighing designs is supposed to be helpful in routine weighing operations to determine weights of (light) objects. Moreover, "the designs are applicable to a great variety of problems of measurement, not only of weights, but of lengths, voltages and resistances, concentrations of chemicals in solutions, in fact any measurements such that the measure of a combination in a known linear function of the separate measures with numerically equal coefficients" (Mood (1946)). Whereas the statisticians have discussed such problems exclusively in the framework of measurements of weights, the general technique seems to have received attention of researchers in other fields as well (Bannerjee (1975), Sinha and Saha (1983)).

In this chapter, we will attempt to provide an up-to-date account of the optimality theory of weighing designs. Of course, we will do so primarily with reference to the problem of measurements of weights.

The formulation to a standard weighing design problem calls for some objects with unknown weights and a weight measuring device which is popularly known as a *balance*. There are two types of balances in use. These are called *chemical* balance and *spring* balance. In a chemical balance, there are two pans and the objects can be placed in either pan. Then standard weights are used to attain *balance*. In a spring balance, there is only one pan and, usually, any number of objects can be placed on the pan. Then the pointer provides a reading which represents the total weight of the objects in the pan (provided the balance has *no* bias). Suppose specifically that there are p objects of true unknown weights $\theta_1, \theta_2, \ldots, \theta_p$ respectively and we wish to estimate them employing N measuring operations using either a chemical balance or a spring balance. Let $y_1, y_2, \ldots, y_N$ denote respectively the recorded observations in these N operations (when *balance* is achieved each time). It is assumed that the observations follow the standard regression model:

$$\mathbf{Y} = \mathbf{X}\boldsymbol{\theta} + \epsilon, \quad E(\epsilon) = \mathbf{0}, \quad E(\epsilon\epsilon') = \sigma^2 \mathbf{I}_N. \tag{8.1.1}$$

Where $\mathbf{X}$ is of order $N \times p$ and is called the weighing *design* matrix. The elements of $\mathbf{X}$ are x_{ij}, $1 \leq i \leq N$, $1 \leq j \leq p$ and a typical element x_{ij} is given by

x_{ij} $= +1$ if the jth object is placed in the right pan during the ith weighing operation

$= -1$ if the jth object is placed in the left pan during the ith weighing operation

$= 0$ if the jth object is *not* utilized in either pan during the ith weighing operation

in the case of a chemical balance;

x_{ij} $= +1$ if the jth object if placed on the pan during the ith weighing operation

$= 0$ if the jth object is *not* utilized during the ith weighing operation

in a spring balance.

Here $\boldsymbol{\theta} = (\theta_1, \theta_2, \ldots, \theta_p)'$ is the vector of true unknown weights (parameters). The vector $\boldsymbol{\epsilon}$ is the so-called vector of (observational) error components satisfying the usual homoscedasticity condition. In the above, we have assumed that neither measuring device has *bias* in the system. We refer to the early works of Hotelling (1944) and Mood (1946) where these concepts have been illustrated with examples. The inference problem centres around estimation of true individual weights of all the objects and, sometimes, also of certain linear combinations of them such as the total weight. The optimality problem arises when we talk about *efficient* estimation in some sense. Surprisingly, efficient estimation of the error variance σ^2 has *never* been an important issue in the context of weighing designs.

We have organized our presentation as follows. In section 2, we present various optimality results for chemical balance designs with or without bias in the measuring device. In section 3, we present the available results for spring balance designs. In either case, we assume that our interest is in the simultaneous estimation of weights of all the objects. In section 4, we address ourselves to the problem of estimation of the total weight. In section 5, we discuss some issues

2. A Study of Chemical Balance Weighing Designs

2.1 Globally Optimal Designs and Hadamard Matrices

We refer to the model (8.1.1) which represents the standard regression model involving the vector $\mathbf{Y}$ of N observations and the vector $\boldsymbol{\theta}$ of true unknown weights $\theta_1, \theta_2, \ldots, \theta_p$ of p objects. The design matrix $\mathbf{X} = ((x_{ij}))$ accounts for the combinations of objects used in the left and right pans of the balance during the different weighing operations. Recall that each x_{ij} assumes values $0, \pm 1$ depending on whether in the ith operation, the jth object has *not* been used at all or has been used in the right pan or in the left pan. The model corresponds to the situation when there is *no* bias in the chemical balance used.

The statistical problem is to estimate the parameter vector θ when the observations undergo the model (8.1.1). The optimality problem is concerned with *efficient* estimation in some sense by a proper choice of the design matrix $\mathbf{X}$ among many designs at our disposal.

The model (8.1.1) is the standard Gauss-Markoff model and the following results are well-known.

Theorem 8.2.1. The parameter vector θ is estimable iff $\text{rank}(\mathbf{X}) = p$ in which case

$$\hat{\theta} = (\mathbf{X}'\mathbf{X})^{-1}\mathbf{X}'\mathbf{Y}, \ \mathbf{V}(\hat{\theta}) = \sigma^2(\mathbf{X}'\mathbf{X})^{-1} \text{ and } \mathbf{I}(\hat{\theta}) = \sigma^{-2}(\mathbf{X}'\mathbf{X}) \tag{8.2.1}$$

where $\hat{\theta}$ is the blue and $\mathbf{V}$ is the dispersion matrix while $\mathbf{I}$ is the information matrix.

Since the elements x_{ij}'s of the design matrix $\mathbf{X}$ assume values $0, \pm 1$, it is clear that $\sum_{j=1}^{p} x_{ij}^2 \leq p$ for all $1 \leq i \leq N$ and $\sum_{i=1}^{N} x_{ij}^2 \leq N$ for all $1 \leq j \leq p$. Let us now denote by $\mathbf{A}$ the matrix $\mathbf{X}'\mathbf{X}$ which is, of course, positive definite since $\mathbf{X}$ has full column-rank. The diagonal elements of the matrix $\mathbf{A}$ are $a_{jj} = \sum_{i=1}^{N} x_{ij}^2 \leq N$, $1 \leq j \leq p$. These algebraic results are helpful in deducing some preliminary but important optimality results first presented by Hotelling (1944). The specific notion of optimality is embodied in the following.

Theorem 8.2.2. Suppose there exists a design matrix $\mathbf{X}_0 = ((x_{0ij}))$ such that $x_{0ij} = \pm 1$ for all $1 \leq i \leq N$, $1 \leq j \leq p$ and, moreover, $\sum_{i=1}^{N} x_{0ij} x_{0ij'} = 0$ for all $1 \leq j \neq j' \leq p$ so that $\mathbf{X}_0'\mathbf{X}_0 = N\mathbf{I}_p$. Then the design matrix $\mathbf{X}_0$ allows for estimation of all the individual weights with the least possible variance for each one of them.

Proof. This proof is due to Moriguti (1954). We have $\mathbf{V}(\hat{\theta}) = \sigma^2 \mathbf{A}^{-1}$ so that $V(\hat{\theta}_j) = \sigma^2 a^{jj}$. Consider spectral decomposition of the matrix $\mathbf{A}$ as $\mathbf{A} = \sum_{l=1}^{p} \alpha_l \xi_l \xi_l'$ where each $\alpha_l > 0$ since $\mathbf{A}$ is p.d.. Then $\mathbf{A}^{-1} = \sum_{l=1}^{p} \alpha_l^{-1} \xi_l \xi_l'$. Thus $a_{jj} = \sum_l \alpha_l \xi_{lj}^2$ and $a^{jj} = \sum_l \alpha_l^{-1} \xi_{lj}^2$. Now we use Cauchy-Schwartz inequality and deduce that $a_{jj} a^{jj} = (\sum_l \alpha_l \xi_{lj}^2)(\sum_l \alpha_l^{-1} \xi_{lj}^2) \geq (\sum_l \xi_{lj}^2)^2 = 1$. This shows that $a^{jj} \geq \frac{1}{a_{jj}}$. Again $a_{jj} = \sum_{i=1}^{N} x_{ij}^2 \leq N$. These together yield $a^{jj} \geq \frac{1}{N}$ and, hence, finally $V(\hat{\theta}_j) \geq \sigma^2/N$ for all $1 \leq j \leq p$ and for *any* design matrix $\mathbf{X}$ of full column-rank.

In other words, the least possible variance for the estimate of any θ_j is σ^2/N. Clearly, $\mathbf{X}_0$ attains this variance for every θ_j. Hence the claim is justified.

Remark 8.2.1. The above result is one of the earliest optimality results known. It shows that a proper choice of the weighing design matrix may enable the experimenter attain maximum precision of $N\sigma^{-2}$ *simultaneously* for the estimates of *each* of the individual weights. We may call such an optimal design *globally optimal* weighing design. Below we give an example to this effect.

Example 8.2.1. $N = 8$, $p = 5$. A globally optimal weighing design exists and is given by

$$\mathbf{X}_0 = \begin{bmatrix} 1 & 1 & 1 & 1 & 1 \\ -1 & 1 & -1 & 1 & -1 \\ 1 & -1 & -1 & 1 & 1 \\ -1 & -1 & 1 & 1 & -1 \\ 1 & 1 & 1 & -1 & -1 \\ -1 & 1 & -1 & -1 & 1 \\ 1 & -1 & -1 & -1 & -1 \\ -1 & -1 & 1 & -1 & 1 \end{bmatrix}$$

when we use a chemical balance without bias.

Such an optimal weighing design is *not* necessarily unique unless $p = N$. Moreover, such a globally optimal design does *not* exist for most combinations of values of (N,p). It is observed that a necessary condition for the existence of a weighing design matrix $\mathbf{X}_0$ (with elements $0, \pm 1$) satisfying the property $\mathbf{X}_0'\mathbf{X}_0 = N\mathbf{I}_p$ is that $N = 0\ (mod\ 4)$ unless $p = 2$ in which case $N = 0\ (mod\ 2)$ is necessary and sufficient. Also for $p \geq 3$, for all *practical* values of $N = 0\ (mod\ 4)$, we can ensure existence of $\mathbf{X}_0$ satisfying the above conditions even though we may *not* be able to do so for every *large* value of $N = 0\ (mod\ 4)$. For $p = 2$, we use $\frac{N}{2}$ times the basic design $\begin{pmatrix} 1 & 1 \\ 1 & -1 \end{pmatrix}$ assuming N to be even. For $p \geq 3$, this is accomplished by using the well-known Hadamard matrices. A Hadamard matrix $\mathbf{H}_N$ of order N is a square matrix of order N with elements ± 1 which satisfies the property $\mathbf{H}_N'\mathbf{H}_N = N\mathbf{I}_N$. A necessary condition for a Hadamard matrix to exist is that $N = 2$ or $N = 0\ (mod\ 4)$ if $N > 2$. Since $p \leq N$ in the set-up of a weighing design providing simultaneous estimation of *all* the individual weights, whenever a Hadamard matrix $\mathbf{H}_N$ is available, we may obtain $\mathbf{X}_0$ by taking *any* $N \times p$ submatrix of $\mathbf{H}_N$. Quite trivially then $\mathbf{X}_0$ will comprise of elements ± 1 and will satisfy the property $\mathbf{X}_0'\mathbf{X}_0 = N\mathbf{I}_p$. Our example above was also based on this principle and we took for $\mathbf{X}_0$ an 8×5 submatrix of $\mathbf{H}_8$. This also explains why $\mathbf{X}_0$ is *not* unique unless $p = N$. Our discussion above regarding availability of globally optimal designs is naturally based on the existing knowledge on the availability of Hadamard matrices. There is a huge literature on this fascinating subject and we will make *no* attempt to discuss it here. We refer to Raghavarao (1971) and Hedayat and Wallis (1978) for various results on this topic. We simply mention in passing that *all* Hadamard matrices up to and including order 100 (and, of course, many more of order exceeding 100) have been made available to the practical experimenters by theoretical researchers in this area. In this context, it may be noted that use of **BIBDs** in constructing Hadamard matrices has also been attempted by some researchers. See Saha and Kageyama (1984) for some results and related references in that area.

We now turn back to the situation when the measuring device has bias in it. This simply means that the pointer in the chemical balance does *not* point to the zero-value in the idle situation (i.e., when no objects are placed in either pan). We will denote by θ_0 the true unknown amount of bias in the balance. Clearly, θ_0

may assume both positive and negative values. Note that the description of the values of x_{ij}'s in the model (8.1.1) is based on the supposition that the standard weights (y_i's) are placed on the left pan. We will assume that the bias component is present in the right (left) pan so that the model expectation of y_i will change from $\sum_{j=1}^{p} x_{ij}\,\theta_j$ to $\theta_0 + \sum_{j=1}^{p} x_{ij}\,\theta_j = \sum_{j=0}^{p} x_{ij}\,\theta_j$ with $x_{i0} = 1, 1 \le i \le N$ and with θ_0 assuming a positive (negative) value. Accordingly, the model (8.1.1) will change to

$$\mathbf{Y} = \mathbf{X}^* \boldsymbol{\theta}^* + \epsilon,\ E(\epsilon) = \mathbf{0},\ E(\epsilon\epsilon') = \sigma^2 \mathbf{I}_N \qquad (8.1.1)'$$

where $\mathbf{X}^* = (\mathbf{1} \mid \mathbf{X}), \boldsymbol{\theta}^* = (\theta_0, \boldsymbol{\theta})'$ are of orders $N\times(p+1)$ and $(p+1)\times 1$ respectively.

Clearly, all the components of $\boldsymbol{\theta}^*$ will be estimable *iff* rank $(\mathbf{X}^*) = p+1$ for which $N \ge p+1$ is necessary. Further, the revised estimates of the components of $\boldsymbol{\theta}$ are obtained from

$$[\mathbf{X}'\mathbf{X} - (\mathbf{X}'\mathbf{1})\,(\mathbf{1}'\mathbf{X})/N]\,\hat{\boldsymbol{\theta}} = \mathbf{X}'\mathbf{Y} - (\mathbf{X}'\mathbf{1})\,(\mathbf{1}'\mathbf{Y})/N$$

resulting in

$$\hat{\boldsymbol{\theta}} = [\mathbf{X}'\mathbf{X} - (\mathbf{X}'\mathbf{1})\,(\mathbf{1}'\mathbf{X})/N]^{-1}\,[\mathbf{X}'\mathbf{Y} - (\mathbf{X}'\mathbf{1})\,(\mathbf{1}'\mathbf{Y})/N] \qquad (8.2.1)'$$

$$\mathbf{I}(\hat{\boldsymbol{\theta}}) = \sigma^{-2}\,[\mathbf{X}'\mathbf{X} - (\mathbf{X}'\mathbf{1})\,(\mathbf{1}'\mathbf{X})/N]$$

Obviously, then, $\mathbf{I}(\hat{\boldsymbol{\theta}}) \le \sigma^{-2}\,\mathbf{X}'\mathbf{X}$ with equality *iff* $\mathbf{X}'\mathbf{1} = \mathbf{0}$.
Therefore, $V(\hat{\theta}_i) \ge \sigma^2/N, 1 \le i \le p$ with equality simultaneously for all $1 \le i \le p$ *iff* $\mathbf{X}'\mathbf{X} = N\mathbf{I}_p$ and $\mathbf{X}'\mathbf{1} = 0$.
We thus conclude that even in the presence of bias, a design matrix $\mathbf{X}_0$ will be globally optimal provided $\mathbf{X}'_0\,\mathbf{X}_0 = N\mathbf{I}_p$ and, moreover, $\mathbf{X}'_0\,\mathbf{1} = \mathbf{0}$. Once more, we can profitably use Hadamard matrices to derive the forms of such globally optimal weighing design matrices. Recall that in such a set-up, $N \ge p+1$. Let $\mathbf{H}_N$ be available. With no loss, the first column of $\mathbf{H}_N$ may be adjusted to $\mathbf{1}$. Leaving this column aside, we can now select any $p\ (\le N-1)$ columns of $\mathbf{H}_N$ to represent the matrix $\mathbf{X}_0$. Clearly, the conditions $\mathbf{X}'_0\,\mathbf{1} = \mathbf{0}, \mathbf{X}'_0\,\mathbf{X}_0 = N\mathbf{I}_p$ are simultaneously satisfied. In Example 8.2.1, the design matrix $\mathbf{X}_0$ thus serves also as a globally optimal weighing design when a chemical balance with bias is in use.

Referring back to the expression for $\mathbf{I}(\hat{\boldsymbol{\theta}})$ in (8.2.1)′, it is seen that whenever the condition $\mathbf{X}'\mathbf{1} = \mathbf{0}$ is satisfied by the design matrix $\mathbf{X}$, the effect of the bias component is immediately eliminated and $\mathbf{I}(\hat{\boldsymbol{\theta}})$ reduces to $\sigma^{-2}\,(\mathbf{X}'\mathbf{X})$ as in (8.2.1). A search for an optimal design in the presence of bias can then be reduced to that without bias provided an optimal design $\mathbf{X}_0$ in the latter situation satisfies $\mathbf{X}'_0\,\mathbf{1} = \mathbf{0}$. This is easily accomplished when the optimal design is based on a Hadamard matrix for which we need to have $N \equiv 0(mod\ 4)$.

Characterization of optimal designs in other cases in the presence of bias appears to be extremely difficult and hence we shall concentrate on the situations where there is no bias. We will present the available results in the following subsections. In some instances, the condition $\mathbf{X'}_0 \mathbf{1} = \mathbf{0}$ will, however, be satisfied and, hence, the result will be applicable to models with bias as well. We will *not* explicitly verify this condition in the different optimality results to be presented below, covering the situations when N does *not* satisfy the condition $N \equiv 0$ (*mod* 4). Clearly, globally optimal designs in the sense described above will no longer be available.

2.2. Extended Universal Optimality for a Full Rank Model

Below we intend to present the available optimality results in a unified way. For this purpose, we recall the general discussion on the choice of optimality criteria as given in section 3 of Chapter One. However, it must be noted that there is a major difference between the earlier set-up (of block designs, row-column designs and the like) and the present set-up (of weighing designs, in particular). This is in the nature of information matrices which are simply non-negative definite (with row and column sums zeroes) in the earlier set-up and strictly positive definite in the present set-up. We now refer to the motivation (regarding the choice of optimality criteria) given in section 3 of Chapter One. All the primary arguments relating to the nature of optimality functionals should apply quite reasonably. Thus, we would expect an optimality functional to satisfy (i), (ii), (iii) and (iv)′. In the present set-up, writing $\mathbf{A}$ for the $p \times p$ p.d. information mtarix underlying p estimable parametric functions, the optimality functionals will be assumed to satisfy

(i) $\Phi(\mathbf{A}) = \Phi(\mathbf{A}_g)$ for every member g of the symmetric group of permutations;

(ii) $\Phi(\mathbf{A}_1) \leq \Phi(\mathbf{A}_2)$ whenever $\mathbf{A}_1 \geq \mathbf{A}_2$;

(iii) $\Phi(\mathbf{A}_1) \leq \Phi(\mathbf{A}_2) \Rightarrow \Phi(t\mathbf{A}_1) \leq \Phi(t\mathbf{A}_2)$ for all positive integers $t \geq 1$;

(iv)′ $\Phi(\sum \mathbf{A}_g) \leq \Phi(p!\mathbf{A})$.

At this stage, we may note that Kiefer (1975) also used the notion of universal optimality in the case of full-rank models. In particular, the optimality functionals were again assumed to satisfy (i), a version of (ii), namely,
(ii)′ $\Phi(tA)$ is non-increasing in t, $t \geq 0$
and the usual condition of convexity. Further, in Proposition 1′ he laid down the same set of sufficient conditions viz., trace maximization and complete symmetry to enable an experimenter to characterize and construct universally optimal designs. However, the statement of the result turned out to be incorrect. (Vide Sinha and Mukerjee (1982)). It is seen that we need an additional condition to be imposed on the class of optimality functionals. This is that
(v) $\Phi(\alpha\mathbf{I}+\beta\mathbf{J}) \geq \Phi(\gamma\mathbf{I})$ whenever $\gamma \geq \alpha+\beta$.
Thus, we come up with the following definition and result.

Definition. A class of optimality functionals Φ is said to define a class of extended universal optimality criteria whenever every member Φ of the said class satisfies the requirements (i), (ii), (iii), (iv)′ and (v).

Theorem 8.2.3. Under a full-rank model, an information matrix $\mathbf{A}_0$ of the form $\mathbf{A}_0 = \gamma\mathbf{I}$ provides extended universal optimality among all competing information matrices whenever $\mathbf{A}_0$ possesses maximum trace.

The proof of this result follows essentially along the same line of arguments as that of Proposition 1 (Restated) given in Section 3, Chapter 1.

Remark 8.2.2. At this stage, we do *not* know of any statistical significance of the condition (v) imposed on the class of optimality criteria. Of course, as has been remarked in Sinha and Mukerjee (1982), the usual optimality criteria e.g., **A**–, **D**–, **E**– and all Φ_p–optimality criteria satisfy (v). Also the generalized optimality criteria and the **MV**–optimality criterion satisfy (v).

Applied to the present set-up of weighing designs, the above theorem demonstrates that for given (N,p), a design matrix $\mathbf{X}_0$ with $\mathbf{A}_0 = N\mathbf{I}_p$ is extended universally optimal in the entire class. Clearly, the proof is based on an easy verification of the fact that $\mathbf{A}_0 = N\mathbf{I}_p$ possesses maximum trace among all information matrices for (chemical balance) weighing designs with design parameters N and p.

Earlier we have seen (Theorem 8.2.2) that designs with information matrix of the form $N\mathbf{I}_p$ are globally optimal. It may be noted that global optimality is *not* based on maximizing a functional defined on the set of information matrices even though every globally optimal design is also extended universally optimal. In any case, weighing designs with the information matrix of the form $N\mathbf{I}_p$ for given N and p thus turn out to be most desirable for simultaneous estimation of all the individual weights with equal importance to each one of them. Once more, we recall that availability of such designs presupposes that the condition $N = 0 \ (mod\ 4)$ for any $p \geq 3$ has been met with in practice. In other situations i.e., when $N \neq 0 \ (mod\ 4)$, we are *not* able to deduce any extended universal or global optimality result unless we restrict to a suitable subclass of possible designs. One such result is reported in Sinha and Mukerjee (1982).

2.3 Generalized Optimal Weighing Designs

Cheng (1980) used the generalized optimality criteria (of type 1) which we have presented in (1.3.1) and (1.3.2). Analogous to Theorem 2.3.2, Cheng (1980) deduced the following result for a full rank model.

Theorem 8.2.4. Suppose in the set-up of a full rank model, $\mathcal{A}$ is the class of all relevant $p \times p$ positive definite information matrices. Suppose moreover that there is an information matrix $\mathbf{A}_0$ (apart from the multiplier σ^{-2}) which has two distinct eigenvalues, with the larger one having multiplicity unity. Then whenever $\mathbf{A}_0$ satisfies

(i) $tr(\mathbf{A}_0) \geq tr(\mathbf{A})$ for all $A \epsilon\ \mathcal{A}$;

(ii) $tr(\mathbf{A}_0^2) < (tr(\mathbf{A}_0))^2/(p-1)$;

(iii) $tr(\mathbf{A}_0) - [p/(p-1)]^{1/2}[tr(\mathbf{A}_0^2) - \dfrac{(tr(\mathbf{A}_0))^2}{p}]^{1/2} \geq$ the corresponding expression involving any other $\mathbf{A} \epsilon\ \mathcal{A}$, the corresponding design $\mathbf{X}_0$ (yielding $\mathbf{A}_0 = \mathbf{X}_0'\mathbf{X}_0$) is optimal w.r.t. any generalized optimality criterion introduced in (1.3.1) and (1.3.2).

By an application of the above theorem, we may now deduce the following optimality result.

Theorem 8.2.5. If $N \geq p$ and $\mathbf{A}_0 = (N-1)\mathbf{I}_p + \mathbf{J}_p$, then the corresponding design $\mathbf{X}_0$ is optimal w.r.t. any generalized optimality criterion among all weighing designs for given N and p.

We omit the proof which involves verification of (ii) and (iii). See Cheng (1980). It can be verified that designs with the above property exist only when $N \equiv 1 \ (mod\ 4)$ if $p \geq 3$. For $N > p \geq 3$, such designs can be easily constructed from a knowledge of $\mathbf{H}_{N-1}$, the Hadamard matrix of order $N-1 \ (\equiv 0 \ (mod\ 4))$. On the other hand, for $N = p \geq 3$, such a design exists only when $p = (1+\alpha^2)/2$ for some odd integer α. An $n \times n$ square matrix with elements ± 1 and with the above property is known as a $\mathbf{P}_n$–matrix so that $\mathbf{P}_n'\mathbf{P}_n = (n-1)\mathbf{I}_n + \mathbf{J}_n$. See Raghavarao (1971) for details on constructional aspects of $\mathbf{P}_n$–matrices.

2.4. Specific Optimality Results

So far the results are quite general in nature and they are valid for the wider class of *all* relevant design matrices in a given situation. In situations where $N \equiv 2 \ (mod\ 4)$ or $N \equiv 3 \ (mod\ 4)$, a good deal of results are yet available even though they apply only in some suitably restricted subclasses of design matrices and/or for some specific optimality criteria.

The subclasses considered are:

$$\mathcal{D}' : \left\{ \mathbf{X} | \mathbf{X} \in \mathcal{D}, \quad \mathbf{X} = ((x_{ij})), \ x_{ij} = \pm 1 \right\}$$

$$\mathcal{D}'' : \left\{ \mathbf{X} | \mathbf{X} \in \mathcal{D}, \quad \mathbf{X} = ((x_{ij})), \ x_{ij} = 0, \pm 1, \ \mathbf{X}'\mathbf{X} = \alpha \mathbf{I}_p + \beta J_p \ \text{for some } \alpha \text{ and } \beta \right\}$$

The restriction to the subclass $\mathcal{D}'$ corresponds to the situation where the experimenter decides to make use of each object in every weighing operation. However, by doing so, the experimeter *cannot* ensure superiority of the *best* (w.r.t. a given criterion) such available design over other designs where *not* all the objects are used in each and every weighing operation. See the discussion in the subsection 2.4.1 below.

The restriction to the subclass $\mathcal{D}''$ does *not* seem to have any *operational* basis or significance. In the early literature available, it has been argued that when all the objects are regarded as being of equal importance, one would like to estimate their weights with equal variance and equal covariance. This is achieved iff the information matrices are completely symmetric and this is precisely ensured by designs in the subclass $\mathcal{D}''$. Again, it is *not* true that the *best* design in $\mathcal{D}''$ is necessarily the best in the wider class of *all* competing designs. See discussion in subsection 2.4.2 below.

The specific optimality criteria studied are for the most part D–optimality. Of late, some results on the E–optimality are also available.

We now present the available results on characterization of optimal designs. We omit the proofs and refer the reader to the papers cited in the text. The constructional aspects of such designs have not been fully investigated in the literature and hence have not been reported here.

2.4.1. Generalized Optimality and the Subclass $\mathcal{D}''$

Sinha and Mukerjee (1982) deduced the following result. Replace condition (v) on the class of optimality functionals by

(v)′ $\Phi(\alpha'\mathbf{I}_p+\beta'\mathbf{J}_p) \geq \Phi(\alpha\mathbf{I}_p+\beta\mathbf{J}_p)$ whenever $\alpha+\beta\geq\alpha'+\beta'$ and $0\leq\beta\leq\beta'$ or $\beta'\leq\beta\leq 0$.

It can be checked that the generalized optimality criteria satisfy (v) ′

Then in the subclass $\mathcal{D}''$, the design matrices

(a) $\mathbf{X}_1^*$, $\mathbf{X}_{-1}^*$ satisfying $\mathbf{X}_{(\pm1)}^{*\prime}\mathbf{X}_{(\pm1)}^* = (N\mp1)\mathbf{I}_p\pm\mathbf{J}_p$ when $N \equiv 1\ (mod\ 4)$

(b) $\mathbf{X}_0^*$, $\mathbf{X}_2^*$, $\mathbf{X}_{-2}^*$ satisfying $\left.\begin{array}{l}\mathbf{X}_{(\pm2)}^{*\prime}\mathbf{X}_{(\pm2)}^* = (N\mp2)\mathbf{I}_p\pm2\mathbf{J}_p\\ \mathbf{X}_0^{*\prime}\mathbf{X}_0^* = (N-1)\mathbf{I}_p\end{array}\right\}$ when $N \equiv 2\ (mod\ 4)$

(c) $\mathbf{X}_0^{**}$, $\mathbf{X}_3^*$, $\mathbf{X}_{-1}^*$ satisfying $\left.\begin{array}{l}\mathbf{X}_{(3)}^{*\prime}\mathbf{X}_{(3)}^* = (N-3)\mathbf{I}_p+3\mathbf{J}_p\\ \mathbf{X}_0^{**\prime}\mathbf{X}_0^{**} = (N-2)\mathbf{I}_p+\mathbf{J}_p\end{array}\right\}$ when $N \equiv 3\ (mod\ 4)$

form a *complete class* w.r.t. any optimality criterion satisfying (i), (ii), (iii), (iv)′ and (v)′. It is then a matter of comparing these designs w.r.t. specific members of the class of generalized optimality criteria. In particular, for $N = p \equiv 2\ (mod\ 4)$, it has been shown that $\mathbf{X}_2^*$ is D–optimal in $\mathcal{D}''$ but $\mathbf{X}_0^*$ is Φ_q–optimal again in $\mathcal{D}''$ for *all* $q\geq0.15$. (Recall (1.3.5) in this context). Thus, $\mathbf{X}_0^*$ is A– and E–optimal in the subclass $\mathcal{D}''$. Jacroux *et al.* (1983c) have shown that when $N \geq p$, $N \equiv 2\ (mod\ 4)$, $\mathbf{X}_0^*$ is uniquely E–optimal in the *entire* class $\mathcal{D}$. This explains why restriction to $\mathcal{D}'$ may *not* be all the time good. A square matrix $\mathbf{X}$ of order $n\times n$ having elements $0, \pm1$ and satisfying the property $\mathbf{X}'\mathbf{X} = (n-1)\mathbf{I}_n$ is known as $\mathbf{S}_n$ matrix. A necessary condition for an $\mathbf{S}_n$ matrix to exist is that $n \equiv 2 (mod\ 4)$. See Raghavarao (1971) for details on constructional aspects of $\mathbf{S}_n$ matrices. Such matrices can be identified with $\mathbf{X}_0^*$ above when $N = p \equiv 2\ (\mathrm{mod}\ 4)$.

2.4.2. Generalized Optimality and the Subclass $\mathcal{D}'$

Jacroux *et al.* (1983c) deduced the following result.

Theorem 8.2.6. For $N \equiv 2\ (mod\ 4)$, a design matrix $\mathbf{X}$ providing

$$\mathbf{X}'\mathbf{X} = \begin{bmatrix}(N-2)\mathbf{I}+2\mathbf{J} & | & 0\\ 0 & | & (N-2)\mathbf{I}+2\mathbf{J}\end{bmatrix}$$

where the orders of the block diagonal matrices differ at most by unity is uniquely optimal in $\mathcal{D}'$ w.r.t. any generalized optimality criterion.

Further, they have also shown that such a design is indeed uniquely D–optimum in the *entire* class $\mathcal{D}$. (This later result is, originally, due to Payne (1974)). This incidentally points out that restriction to the subclass $\mathcal{D}''$ would lead to less efficient design viz., $\mathbf{X}_2^*$ (wrt the D–optimality criterion).

Finally, we may remark that Cheng *et al.* (1985) have provided some further results on Φ_q–optimality (for $0 \le q \le 1$) of X_2^* designs in the unrestricted class for certain combinations of values of (N,p).

2.4.3. D-optimum Designs for $N \equiv 3 (mod\ 4)$

An extensive study of **D**–optimal designs for the case of $N = p \equiv 3\ (mod\ 4)$ was initiated by Ehlich (1964). Subsequently these were extended to the case of $p<N\equiv 3\ (mod\ 4)$ in a series of papers by Galil and Kiefer (1980a, 1980b, 1982a). Contemporarily, Kounias and Chadjipantelis (1983) also worked on this problem and produced results using a different procedure. It is interesting to observe that in the process, various structures of **D**–optimum designs as functions of N and p have been suggested. The general nature of the *information matrix* of an *available* **D**–optimum design for given N and p can be described as follows.

Theorem 8.2.7. For $N \equiv 3\ (mod\ 4)$, a design matrix $\mathbf{X}_0(s_0)$ providing

$$\mathbf{A}_0(s_0) = \begin{bmatrix} \Delta_1(s_0) & -\mathbf{J}(s_0) & \ldots & -\mathbf{J}(s_0) & -\mathbf{J}(s_0) & \ldots & -\mathbf{J}(s_0) \\ & \Delta_1(s_0) & \cdots & -\mathbf{J}(s_0) & -\mathbf{J}(s_0) & \cdots & -\mathbf{J}(s_0) \\ & & \cdots & \Delta_1(s_0) & -\mathbf{J}(s_0) & \cdots & -\mathbf{J}(s_0) \\ & & & & \Delta_2(s_0) & \cdots & -\mathbf{J}_2(s_0) \\ & & & & & \cdots & \Delta_2(s_0) \end{bmatrix}$$

where $\Delta_1(s_0) = (N-3)\mathbf{I}_{[p/s_0]} + 3\mathbf{J}_{[p/s_0]}$, $\Delta_2(s_0) = (N-3)\mathbf{I}_{[p/s_0]+1} + 3\mathbf{J}_{[p/s_0]+1}$, $\mathbf{J}(s_0) = \mathbf{J}_{[p/s_0]}$, $\mathbf{J}_2(s_0) = \mathbf{J}_{[p/s_0]+1}$, $[p/s_0]$ = largest integer $\le p/s_0$, is **D**–optimal in the entire class $\mathcal{D}$ provided $\det\{\mathbf{A}_0(s_0)\} = \max_s \det\{\mathbf{A}(s)\}$ where $\mathbf{A}(s)$ is of the same form as $\mathbf{A}_0(s_0)$ with s_0 replaced by s, $1 \le s \le p$, $s \ne s_0$.

Galil and Kiefer (1982a) have tabulated the values of s_0 for all combinations of (N,p) covering values of $N<100$. Further, they have provided some methods of construction of $\mathbf{X}_0(s_0)$ matrices. Kounias and Chadjipantelis (1983) have developed a different method of producing such **D**–optimal designs. It may be noted that the availability of $\mathbf{X}_0(s_0)$ is *not* ensured for *all* combinations of values of (N,p). Thus for example for $N = 11$, $p = 9$ the maximization of $\det\{\mathbf{A}(s)\}$ yields $s_0 = 7$ but a design matrix $\mathbf{X}_0(s_0)$ does *not* exist. This and other such results have been reported in Galil and Kiefer (1980b). Other related references are Galil and Kiefer (1982b), Moyssiadis and Kounias (1982), Kounias and Farmakis (1984), Farmakis and Kounias (1986) and Chadjipantelis *et al.* (1987). A review article by Raghavarao (1975) includes earlier papers not listed above.

2.4.4. A–optimum Designs for $N \equiv 3\ (mod\ 4)$

We admit at the outset that for historical and technical reasons, the **D**–optimality criterion (for $N \equiv 3\ (mod\ 4)$) has been studied most extensively. The study of **A**–optimality criterion in the wider class of *all* possible design matrices appears to be quite difficult. Wong and Masaro (1984) seem to be the first to make a preliminary study in this direction for $N \le 6$. Subsequently, Cheng *et al.* (1985)

added some general results along the same direction.

We summarize the main result below. Recall the definition of $\mathbf{X}^*_{(-1)}$ in subsection 2.4.1.

Theorem 8.2.8. For each p, there exists a positive integer $N(p)$ such that for all $N \geq N(p)$, $\mathbf{X}^*_{(-1)}$ is A–optimal in $\mathcal{D}$. The integer $N(p)$ may be taken as $N(p) = \max\{(p-2)(p^2-p+16)/8,\ p^2-2\}$.

The implication of this result is that $\mathbf{X}^*_{(-1)}$ is A–optimal in $\mathcal{D}$ for given (N,p) if N is sufficiently large.

At this stage, we may note that in the subclass $\mathcal{D}''$ of relevant designs for given N and p, the competing members are $\mathbf{X}^{**}_0$, $\mathbf{X}^*_{(-1)}$ and $\mathbf{X}^*_{(3)}$ as shown in subsection 2.4.1. Actual computations yield

$$tr(\mathbf{A}_0^{**-1}) = \frac{p-1}{N-2} + \frac{1}{N+p-2}$$

$$tr(\mathbf{A}^{*-1}_{(-1)}) = \frac{p-1}{N+1} + \frac{1}{N-p+1}.$$

It is now easy to verify that for small values of N and p, $tr(\mathbf{A}_0^{**-1})$ can be smaller than $tr(\mathbf{A}^{*-1}_{(-1)})$. One can see this by taking $N = 7, p = 6$. (Both designs $\mathbf{X}^{**}_0$ and $\mathbf{X}^*_{(-1)}$ may be displayed without much difficulty.) Thus it is seen that $\mathbf{X}^*_{(-1)}$ designs are *not* A–optimal even in the restricted subclass $\mathcal{D}''$ for small values of N.

Cheng *et al.* (1985) have furnished some additional results on Φ_q–optimality (for $0 \leq q \leq 1$) of $\mathbf{X}^*_{(-1)}$ designs again for values of N exceeding a certain large number. See also Masaro (1988) for latest results in this direction.

2.4.5. E–optimum Designs for $N \equiv 3 (mod\ 4)$

Recall the designs $\mathbf{X}^*_{(3)}$ and $\mathbf{X}^{**}_0$ introduced in subsection 2.4.1. Cheng (1980) deduced that $\mathbf{X}^*_{(3)}$ is E–optimal in the restricted subclass $\mathcal{D}'$. Later, Jacroux *et al.* (1983c) established that the designs of the type $\mathbf{X}^{**}_0$ are E–optimal in the unrestricted class $\mathcal{D}$ of all designs for given N and p. Further, they also pointed out that an E–optimal design in $\mathcal{D}$ is *not* unique in most cases. We omit the proofs.

2.4.6. MV–optimal Designs

The MV–optimality criterion has been defined in subsection 3.4 of Chapter One in relation to a singular model. For a non-singular model, it seeks to minimize the maximum variance of the estimates of all individual parameters occurring in the model. In the set-up of a weighing design involving p objects, using the notation we have developed so far, an MV–optimal weighing design would seek to minimize $\max_i \{a^{ii}\}$. The following results have been established so far. We assume that N and p are fixed and given.

(i) For $N \equiv 0 (mod\ 4)$, a design with information matrix of the form $N\mathbf{I}_p$ is **MV**–optimal in $\mathcal{D}$. This result is due to Hotelling (1944).

(ii) For $N \equiv 1\ (mod\ 4)$, the design $\mathbf{X}_1^*$ is **MV**–optimal in $\mathcal{D}$. This follows from Cheng (1980). See also Sathe and Shenoy (1986).

(iii) For $N \equiv 2 (mod\ 4)$, the design $\mathbf{X}_0$ mentioned in subsection 2.4.2 is **MV**–optimal in $\mathcal{D}$.

(iv) For $N \equiv 3\ (mod\ 4)$, the design $\mathbf{X}_{(-1)}^*$ is **MV**–optimal in $\mathcal{D}$ *whenever* $N \geq (p-2)(p+1)/2$.

Both the results in (iii) and (iv) have been deduced by Jacroux (1983a). Sharper bounds on N is case of $N \equiv 3(mod\ 4)$ have been obtained by Sathe and Shenoy (1986).

2.5. Chemical Balance Weighing Designs Under a Restricted Set-up

When there is a restriction on the number of objects that can be placed on either pan of a chemical balance, the whole set-up gets altogether changed and the optimality problems get highly complicated. Choice of suitable designs satisfying the symmetry requirement on the part of the information matrix itself poses an interesting problem. Swamy (1982) has furnished some results in this direction. Suppose, specifically, that not more than k_1 (k_2) objects can be placed on the right (left) pan in any weighing operation. We are interested in obtaining suitable weighing designs having this feature and providing completely symmetric information matrices. In this context, it is interesting to note that combinatorial aspects of what are called Balanced Bipartite Weighing Designs (**BBWDs**) have been studied by Huang (1976). Swamy (1982) has successfully applied these **BBWDs** in the present set-up of restricted chemical balance weighing designs and provided preliminary results on designs having completely symmetric information matrices. However, the optimality study seems to be still lacking. Swamy (1981) has also provided some results for efficient estimation of the total weight in a restricted set-up.

3. A Study of Spring Balance Weighing Designs

3.1 Globally Optimal Designs for a Balance With Bias

We refer to the model (8.1.1)$'$. This time $\mathbf{X} = ((x_{ij}))$ where $x_{ij} = 1$ or 0 according as the jth object is or is not utilized in the ith weighing operation, $1 \leq i \leq N, 1 \leq j \leq p$. Moriguti (1954) has shown that the minimum attainable variance for each of the estimated weights in this set-up is $4\sigma^2/N$. A simpler characterization of globally optimal designs (i.e., designs having this minimum variance for estimation of each weight) is obtained as follows.

Theorem 8.3.1. Suppose there exists a design matrix $\mathbf{X}_{00} = ((x_{00ij}))$ such that $\mathbf{X}_{00}'\,\mathbf{1} = (N/2)\,\mathbf{1}$ and $\mathbf{X}_{00}'\,\mathbf{X}_{00} = (N/4)(\mathbf{I} + \mathbf{J})$. Then the design matrix $\mathbf{X}_{00}$ allows for estimation of all the individual weights with the least possible variance of $4\sigma^2/N$ for each one of them.

Proof. Referring to (8.2.1)$'$, based on $\mathbf{X}_{00}$, we have

$$\mathbf{I}(\hat{\theta}) = \sigma^{-2} [\mathbf{X}_{00}' \mathbf{X}_{00} - (\mathbf{X}_{00}'\mathbf{1}) (\mathbf{1}'\mathbf{X}_{00})/N]$$

$$= \sigma^{-2} (N/4) \mathbf{I} \text{ (on simplification)} .$$

Hence the result.

Assuming the existence of $\mathbf{H}_N$ (in case of $N \equiv 0(mod\ 4)$), we can easily construct $\mathbf{X}_0$ based on $\mathbf{H}_N$. We leave aside the column $\mathbf{1}$ in $\mathbf{H}_N$ and pick up any other p columns and, in each, we convert -1's to 0's. The resulting matrix can be easily identified as one having the above properties of $\mathbf{X}_{00}$. This fact was first stated by Mood (1946). It is rather disappointing to note that a globally optimal design is no longer available for a spring balance without bias or for a spring balance with bias in case of $N \not\equiv 0(mod\ 4)$. The above globally optimal design has the added property that the estimates are uncorrelated with each other. Banerjee (1950a) demonstrated that such a property of the estimates may be derived by utilizing the incidence matrices of **BIBDs** in a clever way. Specifically, for given N and p, let $\mathbf{X}$ denote the incidence matrix of a $\mathbf{BIBD}(b = N, v = p, v, r, \lambda)$. Let the parameter values be so chosen that r^2/λ is an integer and it exceeds b. Write $t = r^2/\lambda - N$. Consider then the design matrix

$$\tilde{\mathbf{X}} = \begin{bmatrix} \mathbf{X} \\ \mathbf{0} \end{bmatrix}$$

where $\mathbf{0}$ is of order $t \times p$. In application, this means that the objects are weighed N times according to the design matrix $\mathbf{X}$ and then the bias component only is measured an additional t times. It can be easily verified that $\tilde{\mathbf{X}}'\tilde{\mathbf{X}} - (\tilde{\mathbf{X}}'\mathbf{1})(\mathbf{1}'\tilde{\mathbf{X}})/(N+t) = (r-\lambda)\mathbf{I} + \lambda\mathbf{J} - \frac{r^2}{N+t}\mathbf{J} = (r-\lambda)\mathbf{I} + (\lambda - \frac{r^2}{N+t})\mathbf{J} = (r-\lambda)\mathbf{I}$ since $t = r^2/\lambda - N$. Thus orthogonal estimation of the individual weights is still possible even though this is *not* naturally the most efficient method of estimation. Banerjee (1950a) also extended these results to the case when r^2/λ is *not* an integer but, say, $(r+s)^2/(\lambda+s)$ is an integer for some $s \geq 1$ and also larger than $(N+s)$. In that case we take $\tilde{\mathbf{X}} = \begin{bmatrix} \mathbf{X} \\ \mathbf{J} \\ \mathbf{0} \end{bmatrix}$ where $\mathbf{J}$ is of order $s \times p$ and 0 is of order $t \times p$ with $t = (r+s)^2/(\lambda+s) - (N+s)$.

3.2 Specific Optimality Results for a Balance Without Bias

We admit at the outset that the study of spring balance weighing designs is very much incomplete compared to that of chemical balance wieghing designs. Here the characterization and constructional problems are indeed much more difficult. It appears that after the early work of Mood (1946), there has *not* been any substantial addition of new results in this direction. Specifically, the **D**-optimality criterion has been considered in this study. We present the results below.

Theorem 8.3.2. Consider a spring balance without bias used for measuring p objects in N weighing operations where $N = p \equiv 3\ (mod\ 4)$. Then for any weighing design matrix $\mathbf{X}$,

$$|\mathbf{X}'\mathbf{X}| \leq (N+1)^{N+1}/2^N \tag{8.3.1}$$

where a design matrix $\mathbf{X}_0$ ensuring equality in (8.3.1) is available *iff* $\mathbf{H}_{N+1}$ exists.

Proof: The idea is to establish a link between $\mathbf{X}_0$ and $\mathbf{H}_{N+1}$. Clearly, $\mathbf{X}$ is a non-singular $N \times N$ matrix of 0's and 1's and, hence, $|\mathbf{X}'\mathbf{X}| = |\mathbf{X}|^2$. Consider the enlarged matrix $\tilde{\mathbf{X}}$ defined as

$$\tilde{\mathbf{X}} = \left[\begin{array}{c|c} \mathbf{1} & \begin{matrix} \mathbf{0}' \\ 2\mathbf{X} \end{matrix} \end{array}\right] \tag{8.3.2}$$

so that $\tilde{\mathbf{X}}$ is a square $(N+1)\times(N+1)$ matrix of 0's, 1's and 2's. Obviously, $|\tilde{\mathbf{X}}| = |\mathbf{X}|2^N$. Subtracting the first column of 1's from all other columns in $\tilde{\mathbf{X}}$, we change $\tilde{\mathbf{X}}$ to $\approx{\mathbf{X}}$ is given by

$$\overset{\approx}{\mathbf{X}} = \left[\begin{array}{c|c} \mathbf{1} & \begin{matrix} -\mathbf{1}' \\ ((2x_{ij} - 1)) \end{matrix} \end{array}\right]$$

so that $\overset{\approx}{\mathbf{X}}$ is a square $(N+1)\times(N+1)$ matrix of ± 1's with $|\overset{\approx}{\mathbf{X}}| = 2^N|\mathbf{X}|$.

Now we argue that

$$|\overset{\approx}{\mathbf{X}}{}'\,\overset{\approx}{\mathbf{X}}| \leq \text{product of the diagonal elements of } \overset{\approx}{\mathbf{X}}{}'\,\overset{\approx}{\mathbf{X}} = \prod_{i=1}^{N+1} \left(\sum_{j=1}^{N+1} \overset{\approx}{x}{}_{ij}^2\right) \leq (N+1)^{N+1}.$$

On the other hand, $|\overset{\approx}{\mathbf{X}}{}'\,\overset{\approx}{\mathbf{X}}| = |\overset{\approx}{\mathbf{X}}|^2$ and, hence, finally $|\mathbf{X}'\mathbf{X}| = 2^{-2N}\ |\tilde{\mathbf{X}}\tilde{\mathbf{X}}| = 2^{-2N}\ |\overset{\approx}{\mathbf{X}}{}'\overset{\approx}{\mathbf{X}}| \leq 2^{-2N}(N+1)^{N+1}$ and this establishes (8.3.1). Clearly, whenever $\mathbf{H}_{N+1}$ exists, we can equate $\overset{\approx}{\mathbf{X}}$ to $\mathbf{H}_{N+1}$ and derive the form of $\mathbf{X}_0$ by retracing the above steps. For this, of course, we have to adjust $\mathbf{H}_{N+1}$ in a form which leads to the first column as $\mathbf{1}$ and the first row as $(1\ -1\ -1\ \cdots\ -1)$. This can be trivially achieved and, therefore, $\mathbf{X}_0$ can be formed. On the other hand given a matrix $\mathbf{X}_0$ with $2^N\ |\mathbf{X}_0'\ \mathbf{X}_0| = (N+1)^{N+1}$ one can trace the above steps to construct a square matrix Δ of order $(N+1)\times(N+1)$ having elements ± 1 for which $|\Delta'\Delta| = (N+1)^{N+1}$. Clearly, we can identify Δ with $\mathbf{H}_{N+1}$. Such an $\mathbf{X}_0$ then serves as a **D**-optimal weighing design for $N = p \equiv 3\ (mod\ 4)$ while using a spring balance without bias.

Remark 8.3.1. Mood (1946) used the notation $\mathbf{L}_N$ to denote the matrix $\mathbf{X}_0$ derived above. Banerjee (1948) pointed out that $\mathbf{L}_N$ matrices are indeed a special kind of symmetrical **BIBDs** with parameters $b = v = N, r = k = (N+1)/2$ and $\lambda = (N+1)/4$. It can be easily checked that with the use of $\mathbf{L}_N$ as a design matrix, the variance factors are $V(\hat{\theta}_i) = 4N\sigma^2/(N+1)^2$. Mood (1946) agrued that this is the least attainable variance for $N = p \equiv 3\ (mod\ 4)$.

For our next result, we will assume $N > p$ and further that N admits the representation: $N = \sum_{r \geq 1} n_r \binom{p}{r}$ where n_r's are non-negative integers and $\binom{p}{r} = p!/r!(p-r)!,\ 1 \leq r \leq p$. Denote by $\mathbf{X}^{[r]}$ a matrix of 0's and 1's of order $\binom{p}{r} \times p$ formed of all combinations of p objects taken r at a time.
Also denote by $(\mathbf{X}^{[r]})^{n_r}$ a matrix corresponding to repetition of $\mathbf{X}^{[r]}$ n_r times. Finally, let $\mathbf{U}$ refer to the union (successive application) of the components (design matrices).

Theorem 8.3.3. Suppose $p = 2k-1$ (an odd integer). If $N = \lambda\binom{p}{k}$, then a **D**-optimum design in the subclass of design matrices of the form

$$\mathbf{X} = \mathop{\mathbf{U}}_{r \geq 1} (\mathbf{X}^{[r]})^{n_r}\ ,\ N = \sum_{r \geq 1} n_r \binom{p}{r} \tag{8.3.5}$$

is given by

$$\mathbf{X}_0 = (X^{[k]})^{\lambda}\ . \tag{8.3.6}$$

Proof. It can be easily verified that

$$\mathbf{X}'\mathbf{X} = \sum_{r \geq 1} n_r\,(\mathbf{X}^{[r]\prime}\,\mathbf{X}^{[r]}) = \sum_{r \geq 1} n_r\,(a_r\mathbf{I} + b_r\,\mathbf{J}) = a\mathbf{I} + b\mathbf{J} \tag{8.3.7}$$

where

$$a_r = \binom{p-2}{r-1},\quad b_r = \binom{p-2}{r-2} \tag{8.3.8}$$

$$a = \sum_{r \geq 1} n_r\,a_r,\quad b = \sum_{r \geq 1} n_r\,b_r$$

To characterize a **D**-optimum design, therefore, we have to maximize $(a+pb)a^{p-1}$ for a suitable choice of n_r's subject to the restriction $N = \sum_{r \geq 1} n_r \binom{p}{r}$. Mood (1946) has demonstrated that this maximization is achieved when and only when

$$n_k = \lambda,\ \ n_r = 0 \text{ for } r \neq k \tag{8.3.9}$$

This leads to $\mathbf{X}_0$ in (8.3.6) as the unique **D**-optimum design.

Remark 8.3.2. The use of $\mathbf{X}_0$ leads to the individual variance expressions as $V(\hat{\theta}_i) = 4\sigma^2 p^2/N(p+1)^2$ and Mood (1946) remarked that these are minimum variance factors. There seems to be a typing error in the expression (12) in Mood (1946) which gives the above variance expression as $4\sigma^2 p^2/N(p-1)^2$. Another point to be noted is that in case $p \equiv 3 \ (mod\ 4)$, Mood (1946) asserted that $\mathbf{X}_0$ is **D**-optimal in the entire class of *all* (0,1) matrices. This assertion does not really follow from his arguments even though it is very likely to be true.

Theorem 8.3.4. Suppose $p = 2k$ (an even integer). If $N = \lambda\binom{p+1}{k+1}$, then a **D**-optimum design in the subclass of design matrices $\mathbf{X}$ of the form (8.3.5) is given by

$$\mathbf{X}_0 = (\mathbf{X}^{[k]})^{\lambda} \cup (\mathbf{X}^{[k+1]})^{\lambda} \tag{8.3.10}$$

We refer to Mood (1946) for the proof. It can be seen that in this case the individual variance factors are $V(\hat{\theta}_i) = 4\sigma^2 p/N(p+2)$. However, as observed by Mood (1946), these are *not* minimum variance factors. Using the required number of copies of $\mathbf{X}^{[k]}$ alone (in situations where $N \equiv 0 \ (mod\ \binom{p}{k})$ as well), we can show that $V(\hat{\theta}_i) = 4\sigma^2 \{1+(p-1)^2\}/Np^2$ and this is less than $4\sigma^2 p/N(p+2)$. Mood (1946) remarked that the minimum variance factors could possibly be achieved by a combination of copies of $\mathbf{X}^{[k]}$ and $\mathbf{X}^{[k+1]}$ with more wieghts for the former and less to the latter, instead of equal weights as in (8.3.10).

Remark 8.3.3. For small values of p, Mood (1946) also provided **D**-optimum spring balance weighing designs when the balance has no bias.

Remark 8.3.4. Banerjee (1950b) observed that **D**-optimal designs under the situation $p = 2k - 1$, $N \equiv 0 \ (mod\ \binom{p}{k})$ can also be constructed from an available **BIBD** with parameters $b, v = p, r, k$ for which $b|N$, by taking N/b copies of such a **BIBD**. Thus designs equivalent to $\mathbf{X}_0$ in (8.3.6) may be obtained more simply than using $\mathbf{X}^{[k]}$ as the basic design. For example, for $p = 7, k = 4$, a suitable number of copies of the **SBIBD** $b = v = 7, r = k = 4, \lambda = 2$ may be used as against copies of $\mathbf{X}^{[4]}$ to match the total number of weighing operations (N) which is taken to be a multiple of $\binom{7}{4} = 35$.

We conclude this subsection by stating some further specific optimality results in the subclass analogous to $\mathcal{D}''$ wherein all variances are equal and also all covariances are equal. For given $N = p$, if a spring balance weighing design $\mathbf{X}$ (in the zero bias case) provides estimates with equal variances and covariances, then necessarily $\mathbf{X}'\mathbf{X}$ will have diagonal elements all equal and also off-diagonal elements all equal. This clearly means that $\mathbf{X}$ corresponds to the incidence matrix of a **SBIBD** with $b = v = N = p$, the other parameters $r = k$ and λ being quite arbitrary. It is easy to verify that $\mathbf{X}'\mathbf{X} = (r-\lambda)\,\mathbf{I} + \lambda\mathbf{J}$ so that $|\mathbf{X}'\mathbf{X}| = (r + (v-1)\lambda)\,(r-\lambda)^{v-1} = r^2\,(r - \lambda)^{v-1}$ since the parameter relation $r(k-1) = \lambda(v-1)$ holds in a **BIBD**. Thus, a **D**-optimal design would correspond to a design $\mathbf{X}$ for which $r^2\,(r - \lambda)^{v-1}$ is a maximum. Similarly, it can be concluded that a design would be **A**-optimal if $\frac{1}{r^2} + \frac{v-1}{r-\lambda}$ is minimized while an **E**-optimal design would seek to maximize $(r-\lambda)$. See Raghavarao (1971) for some examples to this effect.

4. Optimum Estimation of Total Weight

The total weight is a naturally suggested linear parametric function and now we will present results pertinent to efficient estimation of it. The key references are Sinha (1972), Dey and Gupta (1977) and Swamy (1980). First we consider the case of a spring balance without bias and with $N = p > 2$.

Theorem 8.4.1. Among all non-singular (0, 1) design matrices of order $N \times N$, the one which minimizes the variance of the estimate of the total weight is given by $\mathbf{X}_0 = \mathbf{J} - \mathbf{I}$.

Proof. Not to obscure the essential steps of reasoning, we will proceed through the following steps.

Step 1. Denote by $\mathbf{X}$ a square non-singular (0, 1) matrix of order $N \times N$. Then the model (8.1.1) applies and we obtain

$$(\sum \hat{\theta}_i) = \mathbf{1}'\hat{\boldsymbol{\theta}} = \mathbf{1}'(\mathbf{X}'\mathbf{X})^{-1}\,\mathbf{X}'\mathbf{Y} = \mathbf{1}'\,\mathbf{X}^{-1}\,\mathbf{Y}$$

$$V(\sum \hat{\theta}_i) = V(\mathbf{1}'\hat{\boldsymbol{\theta}}) = \sigma^2\,\mathbf{1}'(\mathbf{X}'\mathbf{X})^{-1}\,\mathbf{1} \tag{8.4.1}$$

Further, using Cauchy-Schwartz inequality, we obtain

$$\mathbf{1}'(\mathbf{X}'\mathbf{X})^{-1}\,\mathbf{1} \geq (\mathbf{1}'\mathbf{1})^2/\mathbf{1}'(\mathbf{X}'\mathbf{X})\mathbf{1}$$

$$= p^2/\mathbf{1}'(\mathbf{X}'\mathbf{X})\mathbf{1} \tag{8.4.2}$$

Step 2. Consider the subclass of all design matrices $\mathbf{X}$ for which there is at least one 0 in every row of $\mathbf{X}$. Then $\mathbf{1}'(\mathbf{X}'\mathbf{X})\mathbf{1} = \sum_i(\sum_j x_{ij})^2 \leq p(p-1)^2$ so that for such a design matrix $\mathbf{X}$,

$$V(\sum \hat{\theta}_i) \geq \sigma^2 p^2/p(p-1)^2 = \sigma^2 p/(p-1)^2 \tag{8.4.3}$$

Step 3. Using $\mathbf{X}_0 = \mathbf{J} - \mathbf{I}$, we obtain

$$V(\sum \hat{\theta}_i) = V_0\,(\text{say}) = \sigma^2\,\mathbf{1}'(\mathbf{J}-\mathbf{I})^{-1}\,(\mathbf{J}-\mathbf{I})^{-1}\,\mathbf{1}$$

$$= \sigma^2\,\mathbf{1}'(\frac{\mathbf{J}}{p-1} - \mathbf{I})\,(\frac{\mathbf{J}}{p-1} - \mathbf{I})\,\mathbf{1}$$

$$= \sigma^2 p/(p-1)^2 \tag{8.4.4}$$

This shows that $\mathbf{X}_0$ is the best among all competing designs considered in Step 2.

Step 4. The complimentary subclass consists of all such design matrices $\mathbf{X}$ for which one (and only one) row is $\mathbf{1}'$. Since $p = N$, there is *no* error function available and, hence, every estimate is itself the **BLUE** of its expectation. Because there is a row of 1's in $\mathbf{X}$, the corresponding observation is an unbiased estimate of $\sum\theta_i$ and, hence, this serves as its **BLUE** as well. Thus, $V(\sum\hat{\theta}_i) = \sigma^2$ for any such competing design.

It is now trivial to verify that $p/(p-1)^2$ is less than unity whenever $p > 2$. This establishes the result. The above result is due to Sinha (1972).

Remark 8.4.1. For $N = p = 2$, however, the design $\binom{0\ 1}{1\ 0}$ is no better than $\binom{1\ 1}{1\ 0}$ or $\binom{0\ 1}{1\ 1}$ which are found to be the best designs.

Dey and Gupta (1977) extended this result to cover the case of singular weighing designs as well. Precisely, they demonstrated that among all (0,1) design matrices of order $N \times N$ providing unbiased estimation of $\sum\theta_i$, one which minimizes the variance of the estimate of $\sum\theta_i$ is given by $\mathbf{X}_0 = \mathbf{J} - \mathbf{I}$. The interpretation of $\mathbf{X}_0$ is very simple. One has to weigh all the objects skipping one at a time.

Remark 8.4.2. It is interesting to note that the designs which are **D**-optimal for simultaneous estimation of $\theta_1,\ldots,\theta_p$ are no longer optimal for estimation of $\sum\theta_i$. Thus, for example, when $N = p \equiv 3 (mod\ 4)$, an $\mathbf{L}_p$ matrix is **D**-optimal with $V(\hat{\theta}_i) = 4p\sigma^2/(p+1)^2$ and $Cov(\hat{\theta}_i, \hat{\theta}_j) = -4\sigma^2/(p+1)^2, 1 \le i \ne j \le p$. Consequently, for such a design matrix, we observe that $V(\sum\hat{\theta}_i) = 4\sigma^2 p/(p+1)^2 > \sigma^2 p/(p-1)^2$ which was achieved by the use of $\mathbf{J} - \mathbf{I}$. Hence, $\mathbf{L}_p$ is *not* optimal for estimation of $\sum\theta_i$.

We now extend the above result to the case of a chemical balance design without bias again for $N = p > 2$.

Theorem 8.4.2. Among all square non-singular $(0, \pm 1)$ design matrices of order $p \times p, p \ge 3$, the one which minimizes the variance of the estimate of the total weight is given by $\mathbf{X}_0 = \mathbf{J} - \mathbf{I}$.

Proof. To differentiate the chemical balance designs from the spring balance designs, we will assume that in at least one row of such a competing design, both $+1$ and -1 appear together at least once each. In other rows, of course, we may allow any combination of $0, \pm 1$. It is then easy to verify that in such a case, $\mathbf{1}'(\mathbf{X}'\mathbf{X})\mathbf{1} \le (p-2)^2 + (p-1)^3$ (which corresponds to the row totals $(p-2), (p-1),\ldots, (p-1)$). Therefore, analogous to Step 2 of Theorem 8.4.2, we achieve, for any such competing design, $V(\sum\hat{\theta}_i) \ge \sigma^2 p^2/\{(p-2)^2 + (p-1)^3\}$ and this can easily be seen to exceed $\sigma^2 p/(p-1)^2$. The rest of the argument is clear in view of Step 4 discussed above. This establishes the result.

Remark 8.4.3. This result, due to Sinha (1972), indicates that the globally optimal designs are *not* optimal far the estimation of total weight even though they are so for estimation of each of the individual weights. Using the design $\mathbf{H}_N$ (assuming that it exists), we obtain $V(\sum\hat{\theta}_i) = \sum V(\hat{\theta}_i) = \sigma^2 > \sigma^2 p/(p-1)^2$ whenever $p > 2$.

In Sinha (1972), the results were generalized in a different direction w.r.t. spring balance weighing designs. Consider a situation where not more than $k(<p)$ objects can be put together in the balance in any weighing operation. The problem again is to get hold of the most efficient weighing design for estimating the total weight in N weighing operations under this restricted set-up. We omit the derivation of the result which states that any full column-rank (0,1) matrix $\mathbf{X}$ for which $\mathbf{X1} = k\mathbf{1}$ and $\mathbf{X'1} = \frac{Nk}{p}\mathbf{1}$ (Nk/p being assumed to be an integer) provides the most efficient design. Naturally, **BIBDs** may be utilized in such cases. Assuming $N = p$, Sinha (1972) described some non-singular (0,1) matrices $\mathbf{X}$ satisfying the above requirements in both the cases of $(k,p) \equiv 1$ and $(k,p) \equiv \lambda \geq 2$. Dey and Gupta (1977) discussed availability of possibly singular weighing designs in the restricted set-up which are still optimum for estimation of the total weight. Swamy (1980) summarized all these results, discussed the possibility of using **PBIBDs** and presented additional miscellaneous results in this direction. Chacko and Dey (1979) searched for optimal chemical balance weighing designs (in the zero bias case) for estimation of total weight in case $N = p$ from among the designs providing equal variances and equal covariances of the estimates of the individual weights. For $N > p = k$, the best design seems to be $\mathbf{J} - \mathbf{I}$ followed by $\mathbf{J}$ of appropriate orders. Ceranka and Katulska (1986) have discussed the problem of estimation of total weight using singular spring balance weighing designs when the error variances are non-homogeneous.

5. Miscellaneous Topics in Weighing Designs

5.1 Singular Weighing Designs

In practice it is *not* uncommon to find situations where inspite of best efforts on the part of the experimenter, due to unavoidable reasons, one ends up with singular weighing designs. In that case one has to identify the set of estimable parametric functions and act accordingly. If additional weighings can be taken at a subsequent stage, these must be so designed that on the whole estimability is retained to the maximum possible extent and also efficiency can be increased. Raghavarao (1971) and Banerjee (1975) give a fairly complete account of the work done in this area. We will *not* discuss this topic here.

5.2 Weighing Designs Under Autocorrelation of Errors

Banerjee (1965) considered the problem of efficient estimation of the weights in a spring balance assuming an autocorrelation structure for the distribution of the errors. His findings reveal that in situations where the correlation coefficient ρ is positive, **BIBDs** have a better performance than when $\rho = 0$ (uncorrelated errors). The limiting case $\rho \to +1$ is also discussed. See Banerjee (1975) as well.

5.3 Fractional Weighing Designs

It is indeed possible that due to limited resources, an experimenter is unable to conduct a full length experiment for estimation of all individual weights. In that case, a fraction of the design matrix (originally chosen) has to be implemented. Under a nonrandomized (that is, purposive) choice of such a fraction, only some linear functions can be estimated. However, under a randomized (probabilistic) choice of such a fraction, all components of $\underline{\theta}$ might be estimable. Zacks (1966) has studied this problem. Subsequently, Banerjee (1966) has discussed some related results of interest. Later, Beckman (1969) investigated this problem in the framework of spring balance.

5.4 Spring Balance Designs With a String Property

While the statisticians have discussed the weighing problems exclusively in the framework of measurement of weights, the general technique seems to have received attention of researchers in other fields as well. The application of weighing designs to optics seems to have been first pointed out by Marshall and Comisarow (1975) and also independently by Sloane and Harwit (1976). Harwit and Sloane (1979) have explained how the problem of designing Hadamard encoded optical instruments is related to the theory of weighing designs. (See Cheng (1987) in this context). Fulkerson and Gross (1965) and Ryser (1969) have studied some interesting combinational problems involving some classes of (0,1) matrices with what is called the *consecutive* 1's property. Sinha and Saha (1983) call such design matrices as spring balance weighing designs with the *string property*. This basically means that the 1's in every row of the design matrix have to occupy consecutive positions thereby forming *only* one run of 1's. Such (0,1) matrices are found to be useful in optics.

Optimality problems related to (0,1) matrices with the string property were initiated in Sinha and Saha (1983) and followed up in Jacroux (1986), Mukerjee and Huda (1985) and Mukerjee and Saharay (1985). Subsequently, Mukerjee and Huda (1986) have investigated similar problems for (0,1) matrices with *circular string property* (the 1's could form only one run but in a ring instead of in a straight line). This is an interesting area of research where exact optimality results are again hard to establish.

REFERENCES

Banerjee, K. S. (1948). Weighing designs and balanced incomplete blocks. *Ann. Math. Statist., 19*, 394-399.

Banerjee, K. S. (1950a). How balanced incomplete block designs may be made to furnish orthogonal estimates in weighing designs. *Biometrika, 37*, 50-58.

Banerjee, K. S. (1950b). Some contributions to Hotelling's weighing designs. *Sankhyā*, 10, 371-382.

Banerjee, K. S. (1965). On Hotelling's weighing designs under autocorrelation of errors. *Ann. Math. Statist., 36*, 1829-1834.

Banerjee, K. S. (1966). On non-randomized fractional weighing designs. *Ann. Math. Statist., 37*, 1836-1841.

Banerjee, K.S. (1975). *Weighing Designs*. Ed. Marcel Dekker Inc., N. York.

Beckman, R. J. (1969). Randomized spring balance weighing designs. Unpublished Ph.D. dissertation. Kansas State University.

Ceranka, B. and Katulska, K. (1986). Optimum singular spring balance weighing designs with non-homogeneity of the variances of errors for estimation of total weight. *Austr. J. Statist., 28*, 200-205.

Chacko, A. and Dey, A. (1979). Weighing designs optimum for the estimation of total weight. *Sankhyā (B), 41*, 270-276.

Chadjipantelis, T., Kounias, S. and Moyssiados, C. (1987). The maximal determinant of 21 x 21 (0, +1, -1) matrices and D-optimum designs. *J. Statist. Planning and Inference, 16*, 167-178.

Cheng, C. S. (1980). Optimality of some weighing and 2^n fractional factorial designs. *Ann. Statist., 8*, 436-446.

Cheng, C. S., Masaro, J. C. and Wong, C. S. (1985). Optimal weighing designs. *SIAM Jour. Alg. Disc. Meth. 6*, 259-267.

Cheng, C. S. (1987). An application of the Kiefer-Wolfowitz equivalence theorem to a problem in Hadamard transform optics. *Ann. Statist., 15*, 1593-1603.

Dey, A. and Gupta, S. C. (1977). Singular weighing designs and estimation of total weight. *Comm. Statist. (A) Theo. Meth. 6*, 289-295.

Ehlich, H. (1964). Determinantenabeschatzungen für binäre matrizen. *Math. Z. 83*, 123-132.

Farmakis, N. and Kounias, S. (1986). Two new D-optimum designs (83, 56, 12), (83, 55, 12). *J. Statist. Planning and Inference, 15*, 247-258.

Fulkerson, D. R. and Gross, O. A. (1965). Incidence matrices and interval graphs. *Pacific Journal of Mathematics, 15*, 835-855.

Galil, Z. and Kiefer, J. (1980a). D-optimum weighing designs. *Ann. Statist. 8*, 1293-1306.

Galil, Z. and Kiefer, J. (1980b). Optimum Weighing designs. In *Proc. Internat. Conf., Inst. Statist. Math*, Tokyo. 183-189.

Galil, Z. and Kiefer, J. (1982a). Construction methods for D-optimum weighing designs when $n \equiv 3 \ (mod\ 4)$. *Ann. Statist. 10*, 502-510.

Galil, Z. and Kiefer, J. (1982b). On the characterization of D-optimum weighing designs for $n \equiv 3 \ (mod\ 4)$. In *Statistical Decision Theory and Related Topics*, III, *1*, 1-35. Academic Press.

Harwit, M. and Sloane, N. J. A. (1979). *Hadamard Transform Optics*. Academic Press, New York, London.

Hedayat, A. S. and Wallis, W. D. (1978). Hadamard matrices and their applications. *Ann. Statist. 6*, 1184-1238.

Hotelling, H. (1944). Some improvements in weighing and other experimental techniques. *Ann. Math. Statist. 15*, 297-306.

Huang, C. (1976). Balanced bipartite weighing designs. *Jour. Comb. Theo.* (A) *21*, 20-34.

Jacroux, M. (1983a). On MV-optimality of chemical balance weighing designs. *Cal. Statist. Assoc. Bull. 32*, 143-151.

Jacroux, M. and Notz, W. (1983b). On the optimality of spring balance weighing designs. *Ann. Statist. 11*, 970-978.

Jacroux, M., Wong, C. S. and Masaro, J. C. (1983c). On the optimality of chemical balance weighing designs. *J. Statist. Planning and Inference, 8*, 231-240.

Jacroux, M. (1986). On the E-and MV-optimality of spring balance designs with string property. *J. Statist. Planning and Inference, 13*, 89-96.

Kiefer, J. (1975). Construction and optimality of generalized Youden designs. In (J. N. Srivastava ed.) *A Survey of Statistical Designs and Linear Models*. North-Holland, Amsterdam. 333-353.

Kounias, S. and Chadjipantelis, T. (1983). Some D-optimal weighing designs for $n \equiv 3 \pmod 4$. *J. Statist. Planning and Inference, 8*, 117-127.

Kounias, S. and Farmakis, N. (1984). A construction of D-optimal weighing designs when $n \equiv 3 \pmod 4$. *J. Statist. Planning and Inference, 10*, 177-187.

Marshall, A. G. and Comisarow, M. B. (1975). Fourier and Hadamard transform methods in spectroscopy. *Analyt. Chem. 47*, 491A-504A.

Masaro, J. C. (1988). On A-optimal block matrices and weighing designs when $N \equiv 3 \pmod 4$. *J. Statist. Planning and Inference, 18*, 363-376.

Mood, A. M. (1946). On Hotelling's weighing problem. *Ann. Math. Statist. 17*, 432-446.

Moriguti, S. (1954). Optimality of orthogonal designs. *Report of Stat. Appl. Research, Union Jap. Sc. & Eng.* (Tokyo) *3*, 1-24.

Moyssiadis, C. and Kounias, S. (1982). The exact D-optimal first order saturated design with 17 observations. *J. Statist. Planning and Inference, 7*, 13-27.

Mukerjee, R. and Huda, S. (1985). D-optimal statistical designs with restricted string property. *Comm. Statist. (A) Theor. Meth. 14*, 669-677.

Mukerjee, R. and Saharay, R. (1985). Asymptotically optimal weighing designs with string property. *J. Statist. Planning and Inference, 12*, 87-92.

Mukerjee, R. and Huda, S. (1986). Optimal statistical designs with circular string property. *Comm. Stat. (A) Theor. Meth. 15*, 1615-1626.

Payne, S.E. (1974). On maximizing $\det(A^T A)$. *Discrete Math., 10*, 145-158.

Raghavarao, D. (1971). *constructions and combinatorial problems in design of experiments*. Wiley, New York.

Raghavarao, D. (1975). Weighing designs - a review article. *Gujarat Statist. Review, 2*, 1-16.

Ryser, H. J. (1969). Combinatorial configurations. *SIAM J. App. Math. 17*, 593-602.

Saha, G. M. and Kageyama, S. (1984). Balanced arrays and weighing designs. *Austr. Jour. Statist. 26*, 119-124.

Sathe, Y. S. and Shenoy, R. G. (1986). MV-optimal weighing designs. *J. Statist. Planning and Inference, 13*, 31-36.

Sinha, B. K. (1972). Optimum spring balance (weighing) designs. *Proc. All India Convention on Quality and Reliability*. Indian Institute of Technology at Kharagpur.

Sinha, B. K. and Mukerjee, R. (1982). A note on the universal optimality criterion for full rank models. *J. Statist. Planning and Inference, 7*, 97-100.

Sinha, B. K. and Saha, R. (1983). Optimal weighing designs with a string property. *J. Statist. Planning and Inference, 8*, 365-374.

Sloane, N. J. A. and Harwit, M. (1976). Masks for Hadamard transform optics, and weighing designs. *Appl. Opt. 15*, 107-114.

Swamy, M. N. (1980). Optimum spring balance weighing designs for estimating the total weight. *Comm. Statist* (A) *Theo. Meth. 9*, 1185-1190.

Swamy, M. N. (1981). Optimum chemical balance weighing designs for estimating the total weight. *J. Indian Statist. Assoc. 19*, 177-181.

Swamy, M. N. (1982). Use of balanced bipartite weighing designs as chemical balance designs. *Comm. Statist. (A) Theo. Meth. 11*, 769-785.

Wong, C. S. and Masaro, J. C. (1984). A-optimal design matrices $X = ((x_{ij}))_{N \times n}$ with $x_{ij} = 0, \pm 1$. *Lin. and MultiLin. Alg. 15*, 23-46.

Yates, F. (1935). Complex experiments. *J. Roy. Statist. Soc. Suppl. 2*, 181-247.

Zacks, S. (1966). Randomized fractional weighing designs. *Ann. Math. Statist. 37*, 1382-1395.

AUTHOR INDEX

SUBJECT INDEX